T0211357

Introduction to Nonparametric Item Response Theory

MEASUREMENT METHODS FOR THE SOCIAL SCIENCES SERIES

Measurement Methods for the Social Sciences is a series designed to provide professionals and students in the social sciences and education with succinct and illuminating texts on measurement methodology. Beginning with the foundations of measurement theory and encompassing applications on the cutting edge of social science measurement, each volume is expository, limited in its mathematical demands, and designed for self-study as well as formal instruction. Volumes are richly illustrated; each includes exercises with solutions enabling readers to validate their understanding and confirm their learning.

SERIES EDITOR

Richard M. Jaeger, *late of University of North Carolina at Greensboro*

EDITORIAL BOARD

Lloyd Bond, Education, *late of University of North Carolina*
Nancy Cole, *Educational Testing Service*
Bert Green, Psychology, *The Johns Hopkins University*
Ronald K. Hambleton, Education and Psychology, *University of Massachusetts*
Robert L. Linn, Education, *University of Colorado*
Barbara Plake, Buros Institute on Mental Development, *University of Nebraska at Lincoln*
Richard J. Shavelson, *University of California, Santa Barbara*
Lorrie Shepard, Education, *University of Colorado*
Ross Traub, Education, *The Ontario Institute for Studies in Education*

Titles in this series . . .

Introduction to Nonparametric Item Response Theory

volume 5

Klaas Sijtsma
Tilburg University

Ivo W. Molenaar
University of Groningen

SAGE Publications
International Educational and Professional Publisher
Thousand Oaks ▪ London ▪ New Delhi

Copyright © 2002 by Sage Publications, Inc.

All rights reserved. No part of this book may be reproduced or utilized in any form or by any means, electronic or mechanical, including photocopying, recording, or by any information storage and retrieval system, without permission in writing from the publisher.

For information:

Sage Publications, Inc.
2455 Teller Road
Thousand Oaks, California 91320
E-mail: order@sagepub.com

Sage Publications Ltd.
1 Oliver's Yard, 55 City
London EC1Y 1SP

SAGE Publications India Pvt Ltd
B-42 Panchsheel Enclave
PO Box 4109

Library of Congress Cataloging-in-Publication Data

Sijtsma, K. (Klaas) 1955-
    Introduction to nonparametric item response theory/by Klaas Sijtsma
and Ivo W. Molenaar.
        p.   cm. — (Measurement methods for the social sciences; v. 5)
    Includes bibliographical references and index.

    1. Nonparametric statistics. 2. Item response theory—Statistical methods.
    I. Molenaar, Ivo W. II. Title. III. Measurement methods for the social sciences
series; 5.
    QA278.8.S54   2002
    519.5—dc21   2002000393

02   03   04   05   10   9   8   7   6   5   4   3   2   1

Acquisitions Editor:	C. Deborah Laughton
Editorial Assistant:	Veronica K. Novak
Production Editor:	Sanford Robinson
Copy Editor:	Kris Bergstad
Typesetter:	Integra Software Services Pvt, Ltd
Cover Designer:	Sandra Ng

Contents

Preface

This book introduces students and researchers from the social and behavioral sciences to the theory and practice of the highly powerful methods of nonparametric item response theory. Anyone who uses or constructs tests or questionnaires for measuring abilities, achievements, personality traits, attitudes, or opinions may use the methods presented here for obtaining person and item measurements. This includes psychologists, sociologists, political scientists, and educational researchers, but also, for example, medical researchers who study health-related quality of life and marketing researchers who want to gain insight into the preferences and motivations of potential client groups.

Nonparametric item response theory is a family of statistical measurement models that are based on a minimal set of assumptions necessary to obtain useful measurements of persons and items. The models have been formulated for both dichotomous items (e.g., correct/incorrect scoring) and polytomous items (e.g., rating scale scoring). By being defined broadly, they include as special cases most of the well-known parametric item response models for dichotomous items, such as the Rasch model and the 2-parameter and 3-parameter logistic models, and parametric models for polytomous items, such as the partial credit model and the graded response model. The generality of the nonparametric item response models allows them to fit many data sets and still be powerful enough to imply useful measurement properties, such as the ordering of persons using the simple total score (number-correct for dichotomous-item tests, and sum of rating scale scores for polytomous-item tests), and the ordering of the items using the item means.

The book treats nonparametric item response theory at a simple statistical level. Readers are assumed to have followed an introductory statistics course and to be acquainted with concepts such as covariance and correlation, independence, (conditional) probability, and significance testing. Also, knowledge of the principles of measurement through tests and questionnaires and some basic knowledge of classical test theory (observable and true score, measurement error, and reliability and validity) help to grasp fully the theory outlined in this book. The book is intended to teach researchers how to use nonparametric item response theory to analyze their data and build tests and questionnaires. The dominant angle is practical: Many data analysis examples are given, and data analysis is supported by a user-friendly computer program that is used throughout this book for item analysis illustrating the theory of nonparametric item response theory.

We are grateful to Richard M. Jaeger and two reviewers for helping us find the appropriate level of presentation; to Coen A. Bernaards and L. Andries van der Ark for preparing the figures throughout the book; and to Bas T. Hemker, Brian W. Junker, Charles Lewis, Rob R. Meijer, Robert J. Mokken, William F. Stout, L. Andries van der Ark, Wijbrandt van Schuur, and several other colleagues with whom we have done research and published many papers during the past 20 years and who contributed through their work and discussions with us in an almost direct way to this book. The authors, however, remain responsible for its contents.

Klaas Sijtsma and Ivo W. Molenaar

1

Models for Mental Measurement

In the physical sciences, one measures time with a clock, temperature with a thermometer, and length with a ruler. In the social and behavioral sciences, one often works with concepts like intelligence, academic achievement, ego strength, leadership, or personal attitude toward a controversial issue such as euthanasia. The development of measurement instruments for such concepts occurred much later in human history, and their use gives rise to more debate. Why is mental measurement more difficult and more controversial than physical measurement? You are invited to formulate your own answer before reading on.

Systematic study of time, temperature, and length began with the ancient Babylonians and Greeks, and it took several centuries before reliable clocks, thermometers, and rulers were developed. Systematic study of mental measurement was undertaken only a century ago. We believe, however, that this is not just a matter of time lag. Mental measurement is inherently more difficult because the properties being measured do not lend themselves equally easily to direct observation with a method that is simply and universally applicable. It is quite acceptable to say that Harriet is more tolerant about euthanasia than Jane, or that Tom is better in reading than Dick. Daily life manifestations of attitudes, personality traits, and abilities, however, remain less systematic and more variable than those of temperature or length. They are more easily influenced by external circumstances. Moreover, human beings are sometimes inclined to change their behavior for the sake of a more favorable measurement outcome. Examples are cheating on an exam or giving socially desirable answers during an interview.

Assessing mental abilities and mental attributes of individuals may thus be difficult, but at the same time it is highly desirable. Consider

1

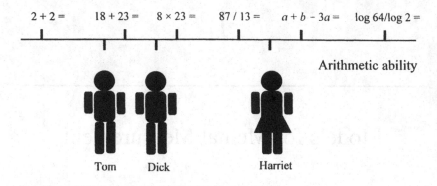

Figure 1.1   Sketch of a Test of Arithmetic Ability.

problems like school admission, personnel selection, or diagnosis of deviant behavior. Decisions in such areas are so important that they should be based on accurate and objective evidence rather than on personal impressions or sloppy measurements. Hiring the wrong person, for example, is not only unfair to the applicants who were not hired but also costly for the employer. Similar aspects play a role in school admission or psychiatric diagnosis: The use of high-quality measurement instruments helps to avoid serious losses for the person being measured as well as for others.

This book shows how a simple statistical model for mental measurement can be used to examine measurement quality and to improve it when that is required. First we present a simplified example that gives the core idea. From Chapter 2 onward, the procedures are systematically discussed and the reader learns how to proceed in more realistic situations.

Suppose we are asked to construct a short test in order to assess the arithmetic ability of Tom, Dick, and Harriet. Not knowing anything about their ages or their skills, we might consider using the six items in Figure 1.1, ranging from very easy to rather difficult. The three respondents are also positioned on the latent continuum of arithmetic ability running from left (low) to right (high).

Our simple measurement model, called the Guttman scalogram, assumes that each item and each person has a location on the latent continuum, and that a correct answer occurs if and only if the person is more able than the item is difficult. Tom knows that $2 + 2 = 4$, but his ability is insufficient to solve $18 + 23$, let alone the more difficult items. Dick's ability exceeds the amount required for $18 + 23$, but he fails on multiplication and division. Harriet is older, or smarter, or both: Can you predict which items she solves and which ones she fails?

In the rest of this book, it is useful to distinguish the actual answer of a respondent from the coding of this answer by the researcher. For example, Tom may have said, or written, that $2 + 2 = 4$; his answer to this item is 4, but the teacher codes it as 1 if it was agreed to use 1 for the correct answer. On the next item, $18 + 23 = ?$, Tom may answer 40, or may answer, "I don't know," which is then coded as 0 if that is the code for an incorrect answer. In another questionnaire, asking for someone's views on euthanasia, agreement with a statement may be coded with 1 and disagreement with 0. These coded answers are called *item scores*. Items with two possible item scores are called *dichotomous* items. Scale analysis and measurement of persons is based on the set of item scores for each person, not on the person's observed actual answers.

If our six arithmetic items are ordered from easy to difficult, then we expect the item score pattern 100000 for Tom, 110000 for Dick, and 111100 for Harriet. In other words, the number of correct answers, also called the total score, also determines which items are correct: Given that Dick has two correct answers, we would be very surprised if these were not the two easiest items. In other tests, however, where the items differ less drastically regarding difficulty, inversions of the kind "an easier item wrong, a more difficult item right" will sometimes occur. Our rule, "a correct answer if person ability exceeds item difficulty," then holds for most person-item pairs, but not for all. This accommodates a bright person being distracted and a dull person having a sudden, bright idea.

The modern models of item response theory (IRT) are thus not deterministic but probabilistic. They say that the probability of a correct answer is high (but not 1) if the person is bright and the item is easy. It is low (but not 0) if the person is dull and the item is difficult. The success probability for a fixed item thus depends on a person's ability or trait level and is called its item response function (IRF); it is usually assumed that this IRF increases if the person has more of the latent ability or the latent trait. In many applications of IRT, it is even assumed that the IRF for each item takes the specific form of a logistic curve, as given in Equations 2.2 and 2.3 in Chapter 2. Then the IRFs of different items differ only in the choice of the parameters of the logistic curve, and we speak of a parametric IRT model.

This book deals primarily with Mokken's model of monotone homogeneity, which essentially poses no other restriction than increasing IRFs and is designed to order persons with respect to a latent trait being measured. We will also meet a second Mokken model, called the double monotonicity model, in which the IRFs not only increase but also must not intersect. This allows us to order the items. Both models are called nonparametric, because any item is free to have an arbitrary IRF, not necessarily given by varying parameters in an algebraic formula like Equations 2.2 and 2.3. This implies that more data sets can be fitted with the Mokken models, and their algorithm for finding a graph like Figure 1.1 is much easier than in the parametric case.

For attitude measurement in particular, but sometimes also for measuring abilities, a more refined scoring per item than just 0 or 1 is found. Answer categories could be labeled *agree, neutral, disagree*, or *always, often, seldom, never*. Such answers can be coded by item scores such as 0, 1, 2, or as 1, 2, 3, 4, respectively. Items with three or more ordered item scores are called *polytomous* items. Later in this book we show that a minor extension of the Mokken models accommodates such data as well. Parametric IRT for polytomous items also exists, but it puts heavier restrictions on the model-data fit and its algorithms are more complicated.

In our example, we presented the location of items and persons as if they were known and then talked about the plausibility of item scores of 0 or 1 given those locations. This is a first step in explaining our model; note that it is quite common in statistics to pretend initially that the unknown parameters are known and to then discuss the probability distribution of outcomes. In reality, we observe the outcomes (say, for 300 persons and 30 items) and then estimate how able each person is and how difficult each item. Moreover, we want to assess whether our model assumptions are realistic, that is, whether there is a good fit between our model and the data. This book presents the tools for such data analyses.

Mental properties are often called "latent," which means hidden. We use the term *latent trait* throughout to indicate any *unobservable* mental property: abilities, achievements, attitudes, personality traits, and so on. This terminology is used because it makes sense to say that one person has more of a latent property than another, but it is impossible to find the position of each person on the latent trait by just one question (like asking someone's age) or one observation (like reading someone's body temperature on a thermometer). Indirectly, however, we can infer people's positions on a latent trait by combining their observable answers to a skillfully chosen set of stimuli or questions. This leaves open how we should infer, how we should combine, and what a skillfully chosen set is.

Measurement models such as IRT models assist researchers in settling these issues. Older models, such as factor analysis and classical test theory, do so by focusing on the outcome for the test as a whole. One major advantage of using IRT is its modeling of specific response probabilities for each specific person-item combination. This allows more flexibility when the test has to be modified or when different versions of a test have been used for different groups. The frequent need to cope with such situations has motivated both major testing agencies and many individual researchers to replace older psychometric models with IRT models.

The construction of a test or scale is typically the joint effort of a domain expert, who constructs the items, and a psychometric expert, who runs a pilot study to check whether the total score on the test (possibly weighted or transformed) can indeed be used to measure people's positions on the latent trait. Skipping the latter step and blindly trusting the domain expert would be unwise. Empirical evidence from many areas has shown that even skilled item writers occasionally include items that are misunderstood by some respondents, items that measure another latent trait than the one intended, or items that are inappropriate for other reasons. Psychometric analysis of pilot results allows us to detect such items, to change or remove them, and to obtain insight into the quality of measurement and the interpretation of test scores.

Such psychometric analyses often make clear why it is recommended that one use an existing measurement instrument whenever possible: It usually comes with instructions for use, and its psychometric properties often can be found in publications. When the instrument is used in a different culture or age group, however, a new psychometric analysis is required because this may affect quality aspects of the test. Empirical research with a deficient test or scale is like visual observation using eyeglasses borrowed from someone else: It is bound to produce unclear or suboptimal results.

At the end of this introduction, we return to our example of measuring arithmetic ability. This time, however, we assume that the test is administered to a class of 30 pupils in fifth grade. Rather than wasting their time on items that are either too easy or too difficult, we administer 10 items of adequate difficulty. Suppose we find that the boys have on average 7.3 items correct and the girls 6.7. May we report that boys calculate better than girls? Note that this implies not only a generalization to other pupils but also to other arithmetic problems: If the gender difference holds only for our 10 items and our 30 pupils, it is hardly worth reporting. The intention to generalize to other persons as well as to other items measuring the same latent trait underlies almost all empirical research with tests or scales, even when this is not made explicit.

Statistics is the tool for such generalizations. The statistical models of IRT are well suited for predictions about other items, other persons, or other test occasions, and for determining whether our measurements are of high enough quality. Mokken's nonparametric IRT models, to which this book is devoted, are simple and widely applicable. They have been successfully applied to the measurement of a wide variety of latent traits in the domains of psychology, sociology, education, political science, medical research, and marketing (some examples are listed in Appendix 1). In the following chapters, we explain how a data analysis using one of the Mokken models proceeds. The acronyms used in this book are listed in Appendix 2.

Exercises

1.1. Specify deficiencies that can be found during actual use of an item but that are hard to predict even if the item writer is a domain expert.

1.2. Which assumptions about the locations of persons and items are characteristic of the Guttman scalogram?

1.3. The argument in favor of a statistical rather than a deterministic IRT model was given in terminology suitable for ability testing (bright, easy, dull, difficult). From the list of applications in Appendix 1, select two cases where this terminology is clearly inappropriate. Invent terms to replace "bright, easy, dull, difficult," and give your argument for why a probabilistic model remains more plausible for the two traits that you have considered.

1.4. The six arithmetic items in Figure 1.1 were presented to Jane. Her six answers were coded as the score pattern 111000. Between which two items would you locate Jane's latent trait value?

1.5. How many different answer patterns can be found if the six arithmetic items of Figure 1.1 form a Guttman scalogram? Also give two examples of answer patterns that would be impossible under this model.

1.6. Mention a benefit of using a statistical item response theory model.

Answers to Exercises

1.1. Misunderstanding by some or all respondents, or measurement of a latent trait other than the one intended.

1.2. Both are located on the same dimension, and a person gives a positive answer if and only if his or her latent trait value lies to the right of the item position.

1.3. Here, too, there are many possible answers. A bright/dull person is, generally, someone who has a high/low position on the latent trait being measured; a difficult/easy item is one for which few/many persons give the positively scored answer. Through personal idiosyncrasies or special circumstances, other characteristics may occasionally produce unexpected answers for person-item combinations in any of the domains listed in the examples; use your experience or your imagination!

1.4. Between 8×23 (third item) and $87/13$ (fourth item).

1.5. Seven permissible patterns, having $s = 0, 1, 2, 3, 4, 5, 6$ item scores of 1, followed by $6 - s$ item scores of 0, respectively. Patterns like 101010 or 000111 would contradict the Guttman model assumption.

1.6. Such models permit us to predict the probability of each answer, given the latent trait value of the person and the parameter(s) of the item.

2

Philosophy and Assumptions Underlying Nonparametric IRT Models for Dichotomous Item Scores

Observable responses to items in a test or questionnaire are assumed to contain information about the latent trait of interest. For personality traits, such as introversion and anxiety, and attitudes, such as the attitude toward abortion and the attitude toward NATO intervention in armed political conflicts in other countries, it may be difficult to construct enough items to obtain sufficient coverage of the latent trait. A problem is that respondents may perceive consecutive statements as repetitions of the same basic question. For example, respondents may feel that a questionnaire measuring the attitude toward abortion asks basically the same question over and over again: Are you in favor, or are you against? This perception may induce halo or horn effects and other unwanted response tendencies, meaning that the respondent does not consider each item separately but responds instead on the basis of a general expectation. For example, someone who considers herself a typical representative of the pro-choice point of view may uncritically answer positively to almost all items in an attitude questionnaire on abortion, without looking very carefully at the exact wording of each item and trying to take a stand on the particular issue raised by that item (halo effect). Someone representing the typical pro-life point of view may do the opposite and respond negatively on the basis of a general conception about the issue (horn effect). For those persons, we will not have valid measures of the trait or attitude. To prevent such problems, careful item and questionnaire construction and clear instructions for filling out the questionnaire are imperative.

For other latent traits, such as the abilities to manipulate fractions in arithmetic or to provide synonyms for given words, examinees have to perform maximally on each separate item, and usually cannot influence their observable performance other than by using their ability. For these ability and achievement tests, it is much easier to construct large numbers of suitable items and, consequently, to obtain different tests that are interchangeable to a high degree. These larger item sets can be stored in *item banks* that also contain background information for each item, for example, about the difficulty, the contents, and the number of times the item has been used before. This information can be used for selecting test versions with a known difficulty and a desired composition by contents. Also, the exposure rate can be controlled; that is, items that have not been used recently might be preferred over items that run the risk of becoming overexposed to a particular group. Selecting tests from item banks is popular and highly useful in educational measurement, where large numbers of examinees are frequently tested and much is at stake.

The nonparametric IRT (NIRT) models discussed in this book can be used to analyze data for the purpose of constructing both personality or attitude scales, and achievement or ability scales. Because NIRT models allow for ordinal measurement, they are well suited for the construction of traditional "stand alone" tests, that is, tests and questionnaires that are presented to each respondent. Typical advanced testing methods that progressively tailor a test to the examinee's ability level during testing, also known as adaptive testing, are handled well using IRT models implying interval scale measurement of both persons and items. Eventually, NIRT also may be capable of such applications, but much work still has to be done to make this possible.

Tests and questionnaires yield measurements on a numerical scale: These are often referred to as scores. There are two types of scores that represent the same principle. The first is the number of items answered correctly, which is typical of many achievement and ability tests. The second is the sum of the scores on the rating scales of a questionnaire, which measures a personality trait or an attitude. In both cases, the score is an unweighted sum of item scores. To enhance interpretation, these scores can be transformed to another scale, such as an IQ scale in intelligence measurement, or educational grading in terms of ordered categories A, B, C, D, and F.

In IRT, observable scores, such as the number-correct, play an important role. For example, the 1-parameter logistic model (1PLM), also known as the Rasch model, transforms the number-correct to the scale of the latent trait, to be denoted θ from now on. Alternatively, the 2-parameter logistic model (2PLM) weights each item score by the discrimination power of the item (to be discussed shortly) and then transforms the sum of the weighted item scores to θ. Thus, the 1PLM considers all items to be equally important for measuring, and the 2PLM gives more weight to better-discriminating items. Likewise, several IRT models for polytomous items transform unweighted or weighted total scores to θ. The IRT models that are central in this book use the unweighted total score for *rank ordering* the respondents on θ.

Parametric and Nonparametric IRT

This book is concerned with IRT models for rank ordering respondents: *nonparametric* IRT models. Models such as the 1PLM and the 2PLM are *parametric* IRT models. To appreciate the difference, we discuss the IRF, which is specified differently for the two types of models, more formally than was done in Chapter 1. In general, the IRF reflects the notion that the higher the latent trait value θ, the higher the probability of correctly answering an item that measures θ. For example, the higher someone's verbal ability, the higher the probability that this person correctly answers an item on a test measuring verbal ability. Thus, the IRF is an increasing function of θ.

In this chapter, we restrict attention to dichotomous items. NIRT for polytomous item scores is discussed in Chapters 7 and 8. Let the random variable X_i denote the item score, which equals 0 for an incorrect answer and 1 for a correct answer. The IRF is defined as

$$P(X_i = 1|\theta) = P_i(\theta), \qquad (2.1)$$

which expresses the dependence of the response probability on θ.

A convenient mathematical choice for the IRF is the logistic function, $\exp(\theta)/[1 + \exp(\theta)]$. This function increases monotonely in θ, first slowly, then quickly, and then slowly again (the function looks like an "S" pushed to the right, and is sometimes called "S-shaped"). The logistic function values vary between 0 and 1, and thus can be interpreted as probabilities. The problem with this function is that it depends only on θ and does not tie the response probability to a particular item, but we all know from experience that some items are more difficult than others, and that this affects the probability of a correct answer. To remedy this problem, we first define a logistic function that depends on the difference between θ and an item's location parameter, δ_i, so that we have an IRF,

$$P_i(\theta) = \frac{\exp(\theta - \delta_i)}{1 + \exp(\theta - \delta_i)}. \tag{2.2}$$

Equation 2.2 is the 1PLM. For the IRF of item i, item parameter δ_i reflects the *location* on θ for which $P_i(\theta) = 0.5$. Because shifting an IRF to the right yields lower response probabilities for all θs, and also a higher δ_i, the location parameter is often interpreted as the *item difficulty*. Figure 2.1 shows five IRFs according to the 1PLM.

The 1PLM assumes that the slopes of all IRFs are the same; see Figure 2.1. In terms of regression analysis, we would say that the relationship between the item score and the latent trait is the same for all items. Obviously, this is a strong assumption, because there seems to be no reason why such relationships should be so restrictive. Alternatively, the 2PLM allows for different slopes. We use notation α_i to denote the slope parameter of item i. The 2PLM is then defined as

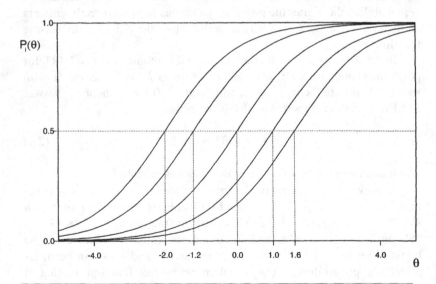

Figure 2.1   Five IRFs With Different Locations (on the θ Axis) According to the 1PLM

$$P_i(\theta) = \frac{\exp[\alpha_i(\theta - \delta_i)]}{1 + \exp[\alpha_i(\theta - \delta_i)]}. \tag{2.3}$$

Item parameter α_i is monotonely related to the slope of the IRF in the point with coordinates (δ_i, 0.5): The steeper the slope, the higher α_i. Compared to a flatter slope, a steeper slope more precisely distinguishes people with low probabilities (and low θs) from people with high probabilities (and high θs), so α_i is interpreted as the discrimination power. Figure 2.2 shows five IRFs with different locations and different slopes.

Equations 2.2 and 2.3 are called parametric functions because they determine the relationship between $P_i(\theta)$ and θ by means of a parametric, in this case logistic, function with scalar parameters δ, and α and δ, respectively. When these parameters are known, the functional relationship is completely determined. Other examples of parametric functions are linear, quadratic, and logarithmic functions. Thus, it should be clear to the reader what makes the 1PLM and the 2PLM parametric IRT models. What is it that makes other IRT models *nonparametric*?

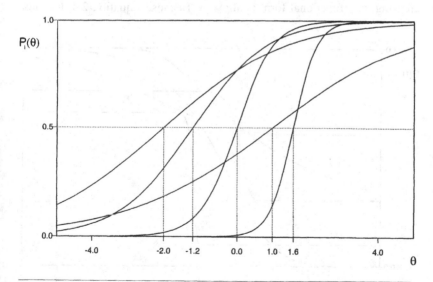

Figure 2.2   Five IRFs With Different Locations (on the θ Axis) and Different Slopes ($\alpha_1 = 0.6$, $\alpha_2 = 1.0$, $\alpha_3 = 2.0$, $\alpha_4 = 0.5$, $\alpha_5 = 3.0$) According to the 2PLM

NIRT models assume that the relationship between $P_i(\theta)$ and θ is governed by *order restrictions*. For example, we assume that for any pair of arbitrarily chosen values θ_a and θ_b, with $\theta_a < \theta_b$,

$$P_i(\theta_a) \leq P_i(\theta_b). \tag{2.4}$$

Equation 2.4 implies that the IRF is a nondecreasing function of θ, and this maintains the interpretation that, for example, the higher the ability for manipulating fractions, the higher the probability of giving the correct answer to an item that measures this ability. When Equation 2.4 is the only restriction on the IRFs, the set of possible IRFs is much greater than the set of IRFs of the 2PLM and, of course, the even more restrictive 1PLM. In fact, Equation 2.3 is just one of the many shapes IRFs can have when restricted only by Equation 2.4.

Figure 2.3 shows examples of IRFs satisfying Equation 2.4. It may be noted that IRFs can be logistic functions as in the 2PLM, but that they can also be partly linear or exponential, that they can have lower (greater than 0) and upper (less than 1) asymptotes, and that they need not be symmetric. Moreover, IRFs may be highly irregular, discrete, or step functions. As long as the relationship between $P_i(\theta)$ and θ is monotonely nondecreasing, any functional form is allowed. Because Equation 2.4 does not

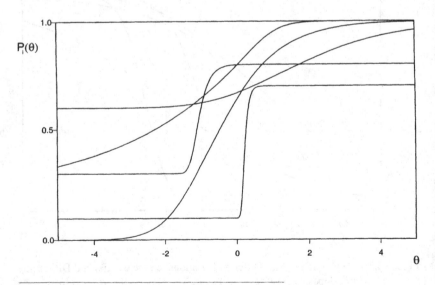

Figure 2.3   Five Monotonely Nondecreasing IRFs

fully determine the IRFs, the definition is nonparametric. Models based on such order-restricted IRFs are nonparametric IRT models. It may be noted that when based on equations such as Equation 2.4, IRT models do not lead to the numerical estimation of interesting parameters such as α_i and δ_i, simply because these parameters are not contained in the definition of the IRFs.

Justification for Nonparametric Item Response Models

Why would we be interested in NIRT models when models such as the 2PLM and software for estimating their parameters are available? The answer is that NIRT models also allow for the estimation of interesting item parameters, for example, the proportion-correct or *P* value indicating difficulty, and scalability coefficients indicating discrimination power. Also, as we will see later on, IRT models assuming only Equation 2.4 for the IRFs imply that the ordering of respondents on the number-correct score reflects, with random error, the ordering on θ.

This all means that if an IRT model is used for constructing a test, and the measurement of respondents on an *ordinal scale* is sufficient for the application envisaged, parametric models might be unduly restrictive for this purpose. Fitting a parametric model with logistic IRFs to the data means the rejection of several items for which the functional relationship with θ cannot be modeled by means of a logistic function. Then, only items with *logistic* IRFs are maintained and the precision of person ordering is based only on these items, whereas precision would be higher if all items that have *nondecreasing* IRFs were included. Some test constructors might argue that in practice, many IRFs are logistic or nearly so. The counterargument is that using the 1PLM means rejecting, for example, all items having IRFs with slopes different from 1; using the 2PLM means rejecting, for example, all items having IRFs with lower asymptotes that are greater than 0 (typical of the multiple-choice item format); and using the 3-parameter logistic model (3PLM; it extends the 2PLM with a lower asymptote or guessing parameter) means rejecting all items with IRFs having upper asymptotes less than 1, as well as all IRFs with an irregular shape, one or more sharp bends, and so on.

Why reject items with nondecreasing IRFs that, in addition to items with logistic IRFs, contribute to precision of person ordering? Our view is that one should not do this when a test is constructed, for example, for selecting the highest-scoring 20 candidates for an expensive training program, the best five candidates for filling in positions in an organization, or the 30% lowest-scoring pupils for remedial teaching. These decisions require a reliable *ordering* of respondents; an NIRT model can be used for this purpose.

We definitely do *not* want to argue for the overall replacement of parametric by nonparametric IRT models. Parametric IRT models lead to point estimates of θ and to interval scales for measuring respondents. Such scales can be very convenient for comparing the results from different tests selected from the same item bank. This technique is known as *equating*. An item bank can also be the basis for tailoring a test to the ability level of an examinee by choosing in consecutive steps items from the bank for which the δ value matches the estimated θ as closely as possible (*adaptive testing*). Parametric IRT models are very powerful methods for applications such as equating and adaptive testing, and their widespread use in many large testing programs corroborates their strength. Nonparametric IRT could be developed for equating and adaptive testing using only ordinal information about person and item measurement (see Laros & Tellegen, 1991, and Huisman & Molenaar, 2001, for first attempts), but compared to parametric IRT, these developments so far have been modest.

The purpose of this book is to explain and demonstrate the theory and application of NIRT. After readers have gone through this and the next six chapters, they will be able to make their own judgments about the usefulness of NIRT. Many real data examples using the computer program MSP for Windows (Molenaar & Sijtsma, 2000) support the theory on NIRT throughout the book.

Models and Data

In what follows, we make a distinction between subgroups of respondents having the same number-correct score and *individuals*. More than statements about the ordering of subgroups on the latent trait θ by means of the number-correct score, statements about the ordering of individuals are subject to *uncertainty*. The reason for this uncertainty is that there is only one measurement value based on a single test or questionnaire administration available for each individual. If we had a large sample of replicated measurements for each individual, we could estimate individual performance very accurately and order individuals with high precision.

Because they are statistical models, IRT models imply statements at the level of subgroups, but measurement practitioners deal with individual test performances. A basic requirement of an IRT model is that it implies the correct ordering of subgroups on θ so that statements about the ordering of individuals on θ are correct except for random measurement error.

More specifically, let the number of items in a test be k and let the unweighted total score be denoted X_+ and defined as

$$X_+ = \sum_{i=1}^{k} X_i. \tag{2.5}$$

As always, X_+ is a fallible score that contains an error component. Thus, it can provide only imperfect information about the ordering on θ. Consider two subgroups with the same X_+ value within groups but different values between groups, say $X_+ = 17$ and $X_+ = 24$, respectively. Under the NIRT models discussed here, the group with the lowest X_+ value has a mean θ smaller than or equal to the mean θ in the other group. Due to measurement error, such group statements may not hold at the individual level. For example, Ann ($X_+ = 24$) may have a lower θ value than Carol ($X_+ = 17$), but she may have been very lucky on the test by guessing correctly at many items, whereas Carol's annoying headache produced unusually many mistakes. An individual's higher θ can thus be obscured by a lower X_+, but for many pairs of individuals, the orderings will agree, in particular when they are sufficiently far apart on the scale. Precision of person ordering can be increased by collecting more information about an individual's trait level within one test administration, that is, by including more items in the test. In practice, however, the number of items is limited, for example, by time constraints and respondents' fatigue and boredom.

When do we know for a particular test or questionnaire that X_+ indeed allows for an ordering (including random errors) of respondents on the latent trait θ? Predictions about the structure of the data matrix can be derived from NIRT models. This predicted data structure can be compared with the actual data. If the predicted and observed structures match, except for differences caused by sample fluctuations, the model is said to explain or fit the data, and the measurement properties of the model, such as ordinal person measurement, are assumed to hold for the test or questionnaire. If the differences between the data structures are too large to be attributed to sample fluctuations, the model cannot explain the data by statistical standards, and the measurement properties of the test or questionnaire cannot be derived from the model. What to do next in situations of model-data misfit is further discussed in Chapters 3, 5, 6, and 8.

Assumptions Underlying NIRT Models

The degree to which a model can explain observed data is determined by the degree of realism of the assumptions about the response processes that lead to an item score. The four assumptions relevant to the two NIRT models for dichotomous items discussed in this book are presented first, followed by a discussion of these two models. Their extension to polytomous items is discussed in Chapter 7.

Unidimensionality

The first assumption is unidimensionality, which means that all items in the test measure the same latent trait. A *psychological* interpretation is that all items measure, for example, the ability to manipulate fractions in arithmetic or tolerance toward euthanasia. The *mathematical* assumption of unidimensionality says that in an IRT model, we need only one latent trait θ to account for the data structure. IRT models assuming unidimensionality are by far the most popular in psychometrics and educational measurement. The main reason for this is that *practical researchers* like their measurement devices to measure one trait at a time, thus facilitating interpretation of test performance. For example, a test measuring a composite of several arithmetic subabilities in combination with spatial orientation and verbal ability yields total scores X_+ that are almost impossible to interpret. That is, a particular score could predominantly reflect a composite of a few arithmetic subabilities and spatial orientation, but also a composite of one particular arithmetic subability and verbal ability. Not only is interpretation dependent on which items a respondent answered correctly, but the comparison of scores of different respondents is also problematic.

Nevertheless, in some cases, latent traits may actually be psychologically complex in the sense that different underlying processes determine the response to a particular item. One example is that the strategy used to solve a particular problem may depend on the developmental stage of the tested child and that consequently different latent traits are needed at different developmental stages. Another example is that some arithmetic items may have a different task composition, triggering varying solution processes between items. In particular, some arithmetic items may be in the form of a small story, thus also requiring verbal ability for their solution, whereas other items are in the form of formal exercises. In such cases, for a wide developmental range or a choice of items that vary highly in task characteristics, a test for these abilities or traits actually might be multidimensional rather than unidimensional. Multidimensional IRT models and estimation methods (e.g., Kelderman & Rijkes, 1994; Reckase, 1997) are available for analyzing the data collected with such tests.

Our point is, however, that whenever possible, one should try to find measurement instruments that measure one meaningful psychological ability or trait at a time. Thus, when some arithmetic items are formulated in such a way that verbal ability is important for solving them, the test constructor should look for alternative ways of presenting the items so that examinees with higher verbal ability no longer have an advantage. That is, a test intended to be an arithmetic test should predominantly measure this ability and no other abilities. We agree with Nunnally (1978, pp. 267–270) that multidimensional tests often have a somewhat coincidental composition, and that for predicting school or job performance, it is more convenient to combine several *unidimensional* tests into a single battery that has high predictive validity.

Local Independence

The assumption of local independence means that an individual's response to item i is not influenced by his or her responses to the other items in the same test. This is a strong assumption. For example, local independence may be violated by learning through practice. This occurs during test administration when the score on the latent trait, for example, of manipulating fractions improves simply because the respondent gains better skills by trying the items. A similar effect can develop during the administration of a personality or attitude questionnaire when the respondent is forced to think about how he or she acts when performing in public, or to express an opinion about an ethical issue such as euthanasia. The exposure to a few well-aimed and crucial statements might direct thinking about one's own behavior, which in turn might influence the responses to the next statements, and the effect could be stronger for respondents who have never contemplated their personality or attitude in such a well-structured environment as a questionnaire. Both examples entail the changing of θ during testing, whereas local independence implies that θ remains the same.

Local independence can be formalized as follows. Let $X = (X_1, X_2, \ldots, X_k)$ be the vector that contains the item score variables, and $x = (x_1, x_2, \ldots, x_k)$ a realization of X. For dichotomous items, each x_i is 0 or 1. The probability of having a score x_i on item i, given a latent trait level θ, is denoted $P(X_i = x_i | \theta)$. Local independence means that

$$P(X = x | \theta) = \prod_{i=1}^{k} P(X_i = x_i | \theta). \tag{2.6}$$

An implication of local independence is that for a fixed value of θ the covariance between items i and j equals 0: $\mathrm{Cov}(X_i, X_j | \theta) = 0$. In a group with varying θ, the covariance is positive, that is, $\mathrm{Cov}(X_i, X_j) > 0$ because the items measure the same θ, but this covariance vanishes for fixed θ because it is completely explained by θ. Zero conditional covariance does not imply local independence (Equation 2.6), however; thus, it is a weaker form of independence. Finally, the reader may have noted that unidimensionality and local independence are related concepts. However, it can be shown that unidimensionality and local independence do not imply one another.

Monotonicity of IRFs

The next assumption is that the conditional probability $P_i(\theta)$ is monotonely nondecreasing in θ (monotonicity assumption). This is Equation 2.4. Obviously, the probability of $X_i = 0$ is also described by the IRF, because

$$P(X_i = 0 | \theta) = 1 - P(X_i = 1 | \theta). \tag{2.7}$$

Figure 2.3 provides examples of IRFs that have the property of monotonicity.

Sometimes the response to an item can be described better by a non-monotone IRF. For example, the statement, "Six months in jail is adequate punishment for the theft of a car" may be endorsed with low probability by both respondents who oppose and those who propose long prison sentences, and with higher probability by those in between. A bell-shaped IRF might adequately describe such a response probability. NIRT models for non-monotone IRFs have been developed but are beyond the scope of this book.

Nonintersecting IRFs

Three assumptions—unidimensionality, local independence, and monotonicity—are enough for many applications of NIRT, in particular when the emphasis is on measuring persons. Some applications, however, require the ordering of the items by difficulty, and then we may add a fourth assumption. This assumption says that the k IRFs are nonintersecting across θ. More specifically, nonintersection means that all IRFs can be ordered and numbered such that

$$P_1(\theta) \leq P_2(\theta) \leq \ldots \leq P_k(\theta), \text{ for all } \theta. \tag{2.8}$$

Equation 2.8 shows that the IRFs do not intersect, and as a consequence, Item 1 is the most difficult item, followed by Item 2, and so on, with the possibility of ties in this difficulty ordering for particular values or intervals of θ. Figure 2.4 provides four nonintersecting IRFs that are also monotonely nondecreasing. The 1PLM is the only well-known parametric IRT model that has nonintersecting IRFs. The practical importance of nonintersecting IRFs is taken up again in the next section and in Chapter 6.

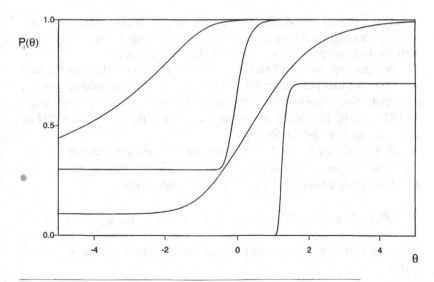

Figure 2.4   Four Nonintersecting IRFs That Are Also Monotonely Nondecreasing

Two Nonparametric Item Response Models

The Monotone Homogeneity Model

The monotone homogeneity model (MHM) is based on the assumptions of unidimensionality, local independence, and monotonicity. The MHM describes item response data that were generated by a set of *homogeneous* (unidimensionality) items having IRFs that are *monotonically* (monotonicity) related to the latent trait.

The practical importance of the MHM is that it implies the ordering of respondents on the θ scale by means of X_+. Thus, the MHM is an IRT model for measuring persons on an ordinal scale. Technically, ordering persons means the following. Let c be an arbitrary fixed value of θ. Then for two fixed values of X_+, denoted s and t, the MHM implies that

$$P(\theta > c | X_+ = s) \le P(\theta > c | X_+ = t), \text{ for } 0 \le s < t \le k. \quad (2.9)$$

Equation 2.9 means that persons can be stochastically ordered on θ by means of X_+. From Equation 2.9 it follows that

$$E(\theta | X_+ = s) \le E(\theta | X_+ = t), \text{ for } 0 \le s < t \le k. \quad (2.10)$$

Equation 2.10 says that within a group with a higher total score t, the mean θ is at least as high as within a group with a lower total score s. This is exactly the ordering property we required of an NIRT model.

We conclude that, for practical purposes, a fitting MHM implies that we can rank order persons on the latent trait θ using their ordering on X_+, even though we cannot obtain numerical estimates of the θs as in parametric IRT models. *This is the key result that justifies the use of the MHM as a measurement model for persons.*

If the roles of θ and X_+ are reversed, the resulting counterparts of Equations 2.9 and 2.10 also hold. That is, for fixed values θ_a and θ_b, and an arbitrary fixed value $X_+ = x_+$, the MHM implies that

$$P(X_+ \ge x_+ | \theta = \theta_a) \le P(X_+ \ge x_+ | \theta = \theta_b), \text{ for } \theta_a < \theta_b, \quad (2.11)$$

and from Equation 2.11, the implication for conditional mean X_+ values is that,

$$E(X_+ | \theta = \theta_a) \le E(X_+ | \theta = \theta_b), \text{ for } \theta_a < \theta_b. \quad (2.12)$$

Equation 2.11 says that X_+ is stochastically ordered by θ, and Equation 2.12 shows that the higher θ is, the higher the expected total score on the test. Although Equations 2.11 and 2.12 are also convenient properties for person measurement, their practical value is limited because the conditioning is on the latent variable, which we cannot observe. The MHM and its practical use in test and questionnaire construction are discussed in Chapters 3, 4, and 5.

The Double Monotonicity Model

The second NIRT model for dichotomous items is the double monotonicity model (DMM). The DMM is based on unidimensionality, local independence, monotonicity, and nonintersection. Because the DMM shares the first three assumptions with the MHM and adds a fourth assumption, it is a special case of the MHM. This means that each data set that can be explained by the DMM can also be explained by the weaker MHM, but the reverse need not be true.

Why would the DMM be an interesting IRT model? To appreciate this, first note that for dichotomous, 0/1 scored items, the expected conditional item score equals the IRF value:

$$E(X_i|\theta) = 0 \times P(X_i = 0|\theta) + 1 \times P(X_i = 1|\theta) = P_i(\theta). \quad (2.13)$$

Next, we replace the conditional probabilities $P_i(\theta)$ in Equation 2.8 (defining nonintersection of k IRFs) by the expected conditional item scores $E(X_i|\theta)$ from Equation 2.13, to obtain

$$E(X_1|\theta) \leq E(X_2|\theta) \leq \ldots \leq E(X_k|\theta), \text{ for each } \theta. \quad (2.14)$$

The item ordering property in Equation 2.14 is called invariant item ordering: The ordering of the items by mean scores is the same, with the exception of possible ties, for each value of θ.

The expected conditional item scores in Equation 2.14 cannot be observed because θ is a latent variable. How do we estimate the item ordering according to Equation 2.14? First, it may be noted that in the population θ is a variable with a density that we denote by $f(\theta)$. Next, we note that the proportion-correct on item i, denoted P_i, is defined by the mean of $E(X_i|\theta) = P_i(\theta)$ (Equation 2.13) taken across $f(\theta)$. Because θ is continuous, we thus have to determine

$$P_i = \int_\theta P_i(\theta) f(\theta) \, d\theta. \quad (2.15)$$

Now, it can be derived, by taking the difference between two integrals, that for two items i and j, with $P_i(\theta) \leq P_j(\theta)$ for all θ, we have

$$P_i - P_j = \int_\theta [P_i(\theta) - P_j(\theta)] f(\theta)\, d\theta \leq 0, \qquad (2.16)$$

or, equivalently,

$$E(X_i|\theta) \leq E(X_j|\theta), \quad \text{for all } \theta \text{ Q } P_i \leq P_j. \qquad (2.17)$$

In other words, an invariant item ordering implies that $P_i \leq P_j$. Also, it may be noted that given an invariant item ordering, $P_i \leq P_j$ gives the ordering of the conditional item means, for all θ; that is,

$$P_i \leq P_j \text{ and invariant item ordering Q } E(X_i|\theta) \leq E(X_j|\theta), \quad \text{for all } \theta.$$
$$(2.18)$$

That is, if we know that $P_i \leq P_j$, and we know that the items have an invariant ordering, as in the DMM, then the item ordering by P values implies the item ordering by expected conditional means. Thus, given that the DMM fits the data, the observable item ordering,

$$P_1 \leq P_2 \leq \ldots \leq P_k, \qquad (2.19)$$

estimates the invariant item ordering by the expected conditional item means; see Equation 2.18. *This is the key result that justifies the use of the DMM as an IRT model for item ordering.* Because the DMM is a special case of the MHM, it also implies person ordering (Equations 2.9 and 2.10).

Finally, it may be noted that given the DMM, the more difficult the item, expressed by its smaller P_i-value, the smaller the probability of having the item correct given any fixed θ (Equation 2.8). In addition to the monotonicity of the IRF, we thus have a second monotonicity assumption here. Hence, the resulting NIRT model is called the *double monotonicity* model.

Next, we say something about the ordering of items in subgroups with varying θs. It may be noted that in Equation 2.16, the distribution of θ, $f(\theta)$, was not restricted to a specific form. Thus, it can have any form without affecting the item ordering, $P_i \leq P_j$. This is an important observation because it implies that we can divide our population of interest into relevant subpopulations, say, boys and girls, that have different θ distributions, say, $f_{\text{boys}}(\theta)$ and $f_{\text{girls}}(\theta)$, and given the conditional item ordering in Equation 2.14

and the nonspecification of $f(\theta)$ in Equation 2.16, the item *ordering* by the P_is in Equation 2.19 for the whole population also is true in each of the subpopulations of boys and girls. Now, it is easy to understand that an invariant item ordering implies the same ordering of the items by P_is, with the exception of possible ties, in each subpopulation of interest. This result takes us to the *practical relevance* of an invariant item ordering.

An invariant item ordering can be of interest in several kinds of research. For example, it may be a sign of differential item functioning or item bias if the ordering of the items by mean or difficulty P_i is not the same in different gender, social, or ethnic groups. Also, if the item ordering in the population is the same for each individual, this helps to interpret the aberrant pattern of item scores produced by a particular examinee. Also, an invariant item ordering (Equation 2.14) or an identical item ordering in subgroups can be relevant for the construction of tests on the basis of psychological theories that predict a particular cumulative structure in the item set, and in test procedures that use starting and stopping rules for different age groups, based on the difficulty ordering of the items. These applications are discussed in more detail in Chapter 6.

Additional Reading

An early review of the MHM and the DMM was given by Mokken (1971) and Mokken and Lewis (1982), and more recent reviews by Mokken (1997), Sijtsma (1998, 2001), Molenaar and Sijtsma (2000), Junker (2001), and Junker and Sijtsma (2001). Also, Holland and Rosenbaum (1986), Holland (1990), Stout (1987, 1990), Junker (1993), and Ellis and Junker (1997) give elaborate mathematical treatments of the MHM and related IRT models. Nonparametric and parametric IRT models are compared at both a theoretical and a practical level by Meijer, Sijtsma, and Smid (1990), and De Koning, Sijtsma, and Hamers (2001). Methods for investigating local independence in real data are discussed by Zhang and Stout (1999) at the mathematical level, and by Douglas, Kim, Habing, and Gao (1998) and Habing (2001) at the practical level. The assumption of monotonely nondecreasing IRFs and methods for investigating this assumption in real data are discussed mathematically by Rosenbaum (1984) and Junker and Sijtsma (2000). Ramsay (1991), Molenaar and Sijtsma (2000), and Douglas and Cohen (2001) discuss these topics from a practical point of view. The assumption of nonintersecting IRFs and methods for investigating nonintersection in real data are discussed and reviewed by Rosenbaum (1987a, 1987b) and Sijtsma and Junker (1996). Stochastic ordering is discussed by Hemker, Sijtsma, Molenaar, and Junker (1997) and Sijtsma and Hemker (2000); also see Grayson (1988) and Huynh (1994). Invariant item ordering is discussed by Sijtsma and Junker (1996, 1997). Finally, Van Schuur (1984), Post (1992), and Post and Snijders (1993) discuss NIRT models for nonmonotone IRFs.

Overviews of parametric IRT are given by Lord (1980), Hambleton and Swaminathan (1985), Hambleton, Swaminathan, and Rogers (1991), and Van der Linden and Hambleton (1997). Fischer and Molenaar (1995) discuss the family of Rasch models, and Boomsma, Van Duijn, and Snijders (2001) give an overview of recent developments in both parametric and nonparametric IRT. Baker (1992) discusses the estimation of parametric IRT models.

This book tries to avoid the mathematical sophistication that is characteristic of much of the NIRT literature. We take the simple MHM and DMM as the two basic models of main interest. The next chapters clarify that these two models are sufficiently developed to justify practical use and that the range of possible applications is broad enough for these two models to be of practical interest for educational and psychological measurement.

Exercises

2.1. The next two cross-tables each contain the frequency counts of 100 respondents for two items i and j. In Table 2.1, all respondents have $\theta = -1$, and in Table 2.2, all respondents have $\theta = 1$.

Table 2.1

$\theta = -1$:	$X_j = 0$	$X_j = 1$	Total
$X_i = 0$	56	24	80
$X_i = 1$	14	6	20
Total	70	30	100

Table 2.2

$\theta = 1$:	$X_j = 0$	$X_j = 1$	Total
$X_i = 0$	4	16	20
$X_i = 1$	16	64	80
Total	20	80	100

a. For each cross-table calculate the covariance, $Cov(X_i, X_j)$. You may use the covariance formula for dichotomous variables:

$$Cov(X_i, X_j) = P_{ij} - P_i P_j, \tag{2.20}$$

where P_{ij} is the population proportion with items i and j both correct, and P_i the population proportion with item i correct.

b. Join the frequencies of both groups into one cross-table and calculate the covariance in the complete group.

c. Which phenomenon did you illustrate in the previous exercises?

d. Calculate the product-moment correlation for both cross-tables. You may use the φ coefficient (this is the product-moment correlation for dichotomous variables):

$$\varphi_{ij} = \frac{ad - bc}{\sqrt{(a + c)(b + d)(a + b)(c + d)}}, \tag{2.21}$$

where a is the frequency in the (0,0) cell of the cross-table, b the (0,1) frequency, c the (1,0) frequency, and d the (1,1) frequency.

e. As in exercise b, but now you have to calculate the φ coefficient.

f. This question goes back to your introductory statistics text book, and you may want to consult it: Can you explain why the covariance and correlation (φ coefficient) yield numerically different results in exercises b and e, and the same results in exercises a and d?

2.2. Under both the MHM and the DMM, the latent trait θ is stochastically ordered by the total score X_+; see Equation 2.9. Suppose the MHM holds for a test consisting of 20 dichotomous items. Do you think that the stochastic ordering property of Equation 2.9 also holds for a simple sum of scores on a subset of, say, 10 items from the same test? Explain your answer.

2.3. Suppose that for two items $P_i = 0.4$ and $P_j = 0.7$.

a. If these two items are part of a test for which the DMM holds, what is the most fundamental statement you can make about the difficulty ordering of the two items?

b. Assume that the DMM holds. Is the item ordering the same for boys and girls? For different grades in school? Explain your answers.

c. Now suppose that the MHM holds, but not the DMM. Can you still calculate P_i and P_j?

d. How would you interpret the difficulty ordering of the two items under the MHM?

2.4. a. Suppose John has probabilities of 0.3, 0.6, and 0.8 for giving the correct answer to Items 1, 2, and 3, respectively. What is his probability of the score pattern (1,1,1)? And (1,0,0)?

b. If the MHM holds, and we know that Ann has a probability of 0.55 and Carol a probability of 0.73 of agreeing with attitude statement i, what can you say about the ordering of Ann and Carol by means of item j from the same scale?

c. If for John $P_i(\theta_{John}) = 0.34$, and for Paul $P_i(\theta_{Paul}) = 0.47$, and item i is part of a test for which the DMM holds, what can you say about these respondents' probabilities on item j from the same test?

d. In Exercises b and c, who have the highest θs?

e. If we actually were to test George and Cynthia, and we knew that $\theta_{George} < \theta_{Cynthia}$, would you be certain that Cynthia will obtain the highest test score X_+? Explain your answer.

Answers to Exercises

2.1. a. $\theta = -1$: $P_{ij} = 0.06$, $P_i = 0.2$, $P_j = 0.3$; $Cov(X_i, X_j) = 0$.
      $\theta = 1$: $P_{ij} = 0.64$, $P_i = 0.8$, $P_j = 0.8$; $Cov(X_i, X_j) = 0$.

   b. Jointly: $P_{ij} = 0.35$, $P_i = 0.5$, $P_j = 0.55$; $Cov(X_i, X_j) = 0.075$.

   c. Local independence (fixed θ); positive covariance across θ.

   d. $\theta = -1$: a = 56, b = 24, c = 14, d = 6; $\varphi_{ij} = 0$.
      $\theta = 1$: a = 4, b = 16, c = 16, d = 64; $\varphi_{ij} = 0$.

   e. Jointly: a = 60, b = 40, c = 30, d = 70; $\varphi_{ij} = 0.302$.

   f. A more common expression for the product-moment correlation is

$$\rho(X_i, X_j) = \rho_{ij} = \frac{Cov(X_i, X_j)}{\sigma(X_i)\,\sigma(X_j)}, \qquad (2.22)$$

where $\sigma(X_i)$ is the standard deviation of X_i; likewise for $\sigma(X_j)$. Because the covariance is in the numerator of the product-moment correlation, $\varphi_{ij} = 0$ if and only if $Cov(X_i, X_j) = 0$ (Exercises a and d). For Exercises b and e, different results for φ_{ij} and $Cov(X_i, X_j)$ were found, because to find the correlation, the covariance has to be divided by the product of the standard deviations; see Equation 2.22. Moreover, for binary items, this product is always smaller than 1 because the maximum of $\sigma(X_i)$ and $\sigma(X_j)$ is 0.5, and their maximum product thus is 0.25.

2.2. To solve this problem, you have to realize that the MHM also holds for any of the $\sum_{k=1}^{19}\binom{20}{k}$ subsets of items. If you leave out items, the items left in the subset remain unidimensional, and their IRFs monotone in θ, and local independence remains valid. This means, that in addition to X_+ for 20 items, all simple sum scores on any subset of items stochastically order θ, and this is true even for scores on individual items. Thus, given the MHM, respondents with $X_i = 0$ have a mean θ that cannot exceed the mean θ of respondents with $X_i = 1$.

2.3. a. Item i is more difficult than item j for each individual of the population of interest (or each θ, if you wish), with the possibility of equal difficulty for some or several, but not all, individuals. This is the property of invariant item ordering.
   b. The item ordering is the same, with the possibility of ties, for all groups you can think of.
   c. You can always calculate them irrespective of the model used for data analysis.
   d. If you know only that the MHM holds, all you can say is that fewer people gave the positive answer to item i.

2.4. a. Due to local independence, $P[(1,1,1)|\theta_{John}] = 0.3 \times 0.6 \times 0.8 = 0.144$; $P[(1,0,0)|\theta_{John}] = 0.3 \times 0.4 \times 0.2 = 0.024$.
   b. Because of monotonicity, Carol can never have a lower probability of agreeing for any item.
   c. Due to nonintersection of IRFs, $P_j(\theta_{John}) \geq 0.34$ and $P_j(\theta_{Paul}) \geq 0.47$ if the IRF of item j is above the IRF of item i, and $P_j(\theta_{John}) \leq 0.34$ and $P_j(\theta_{Paul}) \leq 0.47$ if the IRF of item j is below the IRF of item i. Also, $P_j(\theta_{John}) \leq P_j(\theta_{Paul})$; see Exercise b.
   d. $\theta_{Ann} < \theta_{Carol}$; $\theta_{John} < \theta_{Paul}$; both results due to monotonicity.
   e. Because of Equation 2.12 we would *expect* Cynthia to have the highest X_+ on most occasions. Because of measurement error, however, on a particular test administration, this might not be true.

3

The Monotone Homogeneity Model
Applied to Transitive Reasoning Data

The purpose of this chapter is to illustrate the application of the MHM to a specific set of 12 dichotomously scored items of the 0/1 type. Ten items were intended to measure the ability of transitive reasoning (in accordance with the Piagetian notion; further explained below) of children aged from approximately 7 to 12 years. Two additional items were included for validation purposes. This item set was considered to be a preliminary test for transitive reasoning, and it was hoped that a detailed item analysis would indicate how a final item set should be composed. The purpose of this chapter is to give the reader a first impression of an item analysis according to the MHM. More background on the methods used is provided in Chapter 4.

Transitive Reasoning

The items for transitive reasoning had the following structure. A typical item used three sticks, here denoted A, B, and C, of different length, denoted Y, such that $Y_A < Y_B < Y_C$. The actual test taking had the form of a conversation between experimenter and child in which the sticks were identified by their colors rather than letters. First, sticks A and B were presented to a child, who was allowed to pick them up and compare their lengths, for example, by placing them next to each other on a table. Next, sticks B and C were presented and compared. Then all three sticks were displayed in a random order at large mutual distances so that their length differences were imperceptible, and the child was asked to infer the relation between sticks A and C from his or her knowledge of the relationship in the other two pairs.

All 10 tasks were derived from the literature (see Verweij, Sijtsma, & Koops, 1996); their characteristics are summarized in Table 3.1. Four tasks pertained to inequalities, two tasks to equalities, and four tasks to a mixture of inequalities and equalities. Six tasks had three objects, and the other four tasks had four objects. Five tasks required the comparison of weight, four tasks the comparison of length, and one required the comparison of area. Objects per task could be balls, cubes, discs, sticks, or tubes. Two additional items were included for validation purposes; they had the structure of a task for transitive reasoning, but their solution did not require an inference from knowledge of the relation between objects within pairs. These items will be called pseudo tasks.

Note that the 10 items are not a balanced set in which possibly relevant characteristics are equally represented, as would be desirable for a final test version. They form a typical preliminary test version for which the researcher has relied heavily on what is known from previous research other than test construction. As a result, we would not be surprised to find that this item set is not unidimensional and the quality of the items is not uniformly high.

Purposes of Item Analysis by Means of the Monotone Homogeneity Model

The purposes of an item analysis of the transitive reasoning data by means of the MHM are to obtain evidence that the test measures transitive reasoning and no other latent traits (unidimensionality), and that respondents can be rank ordered with respect to transitive reasoning ability (technically, denoted θ) by means of the number of correctly answered transitive reasoning items (total test score, denoted X_+; see Equation 2.5) in the sense of a stochastic ordering (Equation 2.9). Statistical methods are used to obtain this evidence by investigating whether the assumptions of the MHM are tenable given the structure of the data. The ideal result would be that the model fits the data within the limits of sampling error, and that the test thus allows for unidimensional ordinal person measurement. Results from data analyses are rarely like that. For example, some items may tap other latent traits in addition to transitive reasoning, the overall difficulty level may be too low or too high, and reliability of X_+ may be

Table 3.1 Description of Ten Transitivity Tasks and Two Pseudo Tasks. Item Mean Scores (P_i) for Total Group and Separate Grades; Group Size Under Group Number.

Task	Property	Format	Objects	Measures	Total Group 2–6 $n = 425$	Grades 2 $n = 86$	3 $n = 85$	4 $n = 82$	5 $n = 85$	6 $n = 87$
9	length	$Y_A = Y_B < Y_C = Y_D$	Sticks	12.5, 12.5, 13, 13 (cm)	.30	.21	.33	.30	.26	.40
12	pseudo	$Y_A = Y_B < Y_C = Y_D$			.48	.66	.51	.39	.52	.30
10	weight	$Y_A = Y_B < Y_C = Y_D$	Balls	60, 60, 100, 100 (g)	.52	.41	.58	.39	.55	.67
11	pseudo				.64	.49	.60	.67	.71	.75
4	weight	$Y_A = Y_B = Y_C = Y_D$	Cubes	65 (g)	.78	.84	.65	.82	.82	.79
5	weight	$Y_A < Y_B = Y_C$	Balls	40, 50, 70 (g)	.80	.71	.80	.78	.82	.90
2	length	$Y_A = Y_B = Y_C = Y_D$	Tubes	12 (cm)	.81	.81	.73	.89	.79	.83
7	length	$Y_A > Y_B = Y_C$	Sticks	28.5, 27.5, 27.5 (cm)	.84	.72	.84	.87	.87	.93
3	weight	$Y_A > Y_B > Y_C$	Tubes	45, 25, 18 (g)	.88	.80	.93	.91	.85	.93
1	length	$Y_A > Y_B > Y_C$	Sticks	12, 11.5, 11 (cm)	.94	.85	.99	.98	.93	.97
8	weight	$Y_A = Y_B > Y_C$	Balls	65, 40, 40 (g)	.97	.93	.96	.98	.98	.99
6	area	$Y_A > Y_B > Y_C$	Discs	7.5, 7, 6.5 (diameter; cm)	.97	.95	.95	.99	.98	1.00

33

too low. The item set analyzed here initially had several shortcomings that we believe add to the didactical value of this example.

Data and Sample

We analyzed data from a sample of 425 pupils aged from approximately 7 to 12 years. These pupils attended Grades 2 through 6 in 10 Dutch primary schools. Grades and genders were equally represented in the sample. Children were tested individually. A correct answer was scored 1, and an incorrect answer 0.

Strategy of Item Analysis, and Software

We analyze the transitive reasoning data following a stepwise procedure and discuss the results with sufficient detail and structure for the reader to gain insight into an item analysis using the MHM. It may be noted that each data set has its own unique characteristics that may call for somewhat different strategies of analyzing the data. Analyses of different data sets share sufficiently many similarities, however, to warrant a stepwise, structured approach similar to the one followed here.

In the present example, we distinguish four item analysis rounds. The first is aimed primarily at getting to know the data set and identifying the worst-fitting items using relatively simple scalability coefficients that tell us the degree to which an item fits into a test or a scale. These items are deleted in the second round, and a method for estimating the IRFs is used in combination with the scalability coefficients from the first analysis to identify other nonfitting items. This process is repeated in the third and fourth item analysis rounds until a satisfactory end result is obtained.

The computations were done with the computer program *Mokken Scale analysis for Polytomous items* (acronym MSP). MSP was especially designed for item analysis using nonparametric IRT models such as the MHM. MSP can handle dichotomous item scores as well as polytomous item scores of the Likert type, which are often used in personality inventories and attitude questionnaires. The latest version of MSP (Molenaar & Sijtsma, 2000) is a Windows program that gives the user statistical results at several levels of detail and also has graphical facilities. We used MSP throughout for item analysis. Appendix 3 informs the reader how to obtain a copy of MSP and its manual.

Analysis of Transitive Reasoning Data Using the Monotone Homogeneity Model

Item Analysis 1: Analysis of All Twelve Items

Step 1.1: Inspecting the Item Popularities (P_i Values). Before actually starting an item analysis using the MHM, do a first inspection of the mean scores of the items (the P_is; $i = 1, 2, \ldots, 12$; see Equation 2.15). Table 3.1 shows that in the total group, the means are higher than 0.9 for three items and higher than 0.8 for seven items. This is an indication that several tasks may be too easy for pupils from the higher grades; this is further discussed below for the individual grades.

The reader may note that in Table 3.1, items are arranged according to increasing P_is, but the numbering is that which was originally chosen by the researcher when entering the items in his data file. Of course, one might prefer to renumber the items according to increasing P_i, such that Item 1 was the most difficult (lowest P_i value), and so on, but when leaving out items at a later stage of item analysis, this would call for other renumberings and probably much confusion due to the changing identification numbers of fixed items across analyses. Instead, we chose to stick to the item numbers provided by the researcher throughout all the analyses. The reader may feel that this issue is more an aside than a main point, but our experience is that it is potentially confusing to inexperienced researchers, and for that reason, we mention it at the beginning of the first data analysis.

The results for separate grades, also given in Table 3.1, show that there is a tendency for the P_i values of several items to increase with grade, but that this is probably not true for Items 4 and 2. For Item 2, the proportion correct in Grade 6 is almost the same as in Grade 2, and for Item 4, it is even lower in Grade 6. These results can be taken as the first warning to be suspicious of Items 2 and 4. For Items 9, 5, 3, and 1 (mentioned in the order in which they appear in Table 3.1), the trend of increasing proportion correct is not always obvious, but deviations are small, and the proportion correct in Grade 6 is always higher than in Grade 2. The results for grades also show that some of the items may be too easy not only for the highest grade but for lower grades as well. Finally, pseudo task 11 shows an increasing trend across grades. The trend for pseudo task 12 is remarkably jumpy, which indicates that we should be suspicious of this task in further analyses.

Step 1.2: Inspecting the Sign of the Inter-Item Covariances. An item analysis according to the MHM typically starts with an inspection of the sign of the covariances for all pairs of items, which must be nonnegative under the MHM; see Chapter 4, Theorem 4.1. For 12 items, the number of covariances is $\binom{12}{2} = 66$, of which 22 were negative. Thus, the MHM does not hold for all 12 items. All items were involved in negative covariances with other items. The two pseudo tasks had the most negative covariances with the other items, but several other items also had negative covariances with each other. Thus, based on covariances alone, it was not unequivocally clear which items were likely candidates for removal from the scale. As we will see shortly, better methods are available for this purpose.

Step 1.3: Inspecting the Scalability of Individual Items. Given the results so far, at this stage of the data analysis a useful strategy may be to study the characteristics of the individual items in an effort to explore the structure of the data for the 12 items. A fruitful tool for this investigation is the scalability coefficient for individual items, denoted H_i ($i = 1, 2, \ldots,$ 12). The H_i coefficient, explained briefly below and in more detail in Chapter 4, indicates how well item i fits with the other items for the purpose of ordering respondents.

Theoretical Sidestep: The Item Scalability Coefficient H_i. For an arbitrary collection of items, not necessarily in agreement with the MHM, the H_i coefficient can attain negative and positive values. Its theoretical minimum is negative and depends on characteristics of the test and the distribution of θ. Its maximum always equals 1. The MHM restricts the possible values of H_i. Given that the MHM holds, it can be shown that $0 \le H_i \le 1$ (see Chapter 4, Theorem 4.3). Thus, from a *theoretical* point of view, under the MHM, H_i must be nonnegative, and any item with a nonnegative H_i value is acceptable. It may be noted that under the MHM, items with small positive H_i values, say, ranging from 0 to 0.3, have positive but weak discrimination power. In *practice*, such items are not very useful in a test because they contribute very little to a reliable person ordering.

Now, we look at the reverse situation, which is the situation researchers typically face: We have a data set and want to know whether the MHM fits these data. Because the MHM *implies* nonnegative H_is, we know that negative values are in conflict with the model. Also, items with H_i values, say, between 0 and 0.3, tend to have weak discrimination

power, and sometimes their IRF *decreases* across small intervals of the
θ scale. Because such items are not very useful in a test, a rule of thumb is
that H_i values less than 0.3 are considered unacceptable for an item to be
admitted to the test.

For each H_i calculated from the sample, a standard normal deviate Z_i
is computed that can be used to test one-sidedly the null hypothesis that
$H_i = 0$ in the population against the alternative that $H_i > 0$. It can thus be
decided whether there is some evidence that an item has at least positive
discrimination ($H_i > 0$) or not ($H_i = 0$). This is the standard statistical test
for H_i coefficients.

In special cases, however, one may also test the null hypothesis of
$H_i = 0$ against the alternative that $H_i < 0$. This can be done in situations
where a negative H_i is unexpected for a particular item and the
researcher is reluctant to reject the item from the test. A significant
result at least rules out, with a probability of being wrong equal to the
Type I error, the possibility that the data for this item can be explained
by the MHM, and thus provides the researcher with more convincing
evidence that his or her expectations might have been wrong.

Step 1.3 Continued. If we were to blindly follow our rule of thumb and
accept only items with $H_i > 0.3$, all items except Item 6 would be candidates for removal from the test (Table 3.2, First Analysis). This naive
strategy would ignore the likely event that a few poorly fitting items might
negatively influence the quality indices of the other items. A more con-
structive strategy is to move forward, excluding items one at a time or in
pairs, and then to check at each step whether the quality of the remaining
item set as a test has improved. A part of this strategy should be to use the
knowledge of the item content in combination with numerical information
and, for example, to remove an item only if it is understood why it dys-
functions, by a priori or ad hoc interpretation. Following this strategy, the
two items most likely to be removed are the pseudo tasks 11 and 12:
These items have negative H_i values, which are inadmissible on theoretical grounds under the MHM, and moreover, they were expected a priori
not to fit in with the other items, because their solution does not require
transitive reasoning.

This concludes Item Analysis 1. We continue with the second item
analysis and again distinguish several analysis steps: The first two were
encountered in the first item analysis round (inspection of inter-item
covariances and item scalability coefficients), and the other two are new
(scalability of the total item set and estimating the IRFs).

Table 3.2 Mean Scores, Scalability Coefficients, and Standard Normal Deviates for 12 Items and the Total Score; for Four Data Analyses.

Task	Property	Format	First Analysis $k = 12$		Second Analysis $k = 10$		Third Analysis $k = 8$		Fourth Analysis $k = 7$	
			H_i	Z_i	H_i	Z_i	H_i	Z_i	H_i	Z_i
9	length	$Y_A = Y_B < Y_C = Y_D$	.17	4.36	.29	5.55	.44	7.23	.50	7.55
12	pseudo		-.14	-4.65						
10	weight	$Y_A = Y_B < Y_C = Y_D$	.14	4.95	.28	7.51	.46	10.09	.52	9.91
11	pseudo		-.03	-0.97						
4	weight	$Y_A = Y_B = Y_C = Y_D$	.05	1.95	.04	1.35				
5	weight	$Y_A < Y_B < Y_C$	.09	3.75	.14	5.29	.20	5.34		
2	length	$Y_A = Y_B = Y_C = Y_D$	.08	3.31	.07	2.73				
7	length	$Y_A > Y_B = Y_C$	.18	6.87	.25	9.19	.39	10.81	.51	11.37
3	weight	$Y_A > Y_B > Y_C$	.19	6.58	.29	9.92	.45	12.12	.53	12.09
1	length	$Y_A > Y_B > Y_C$	.21	5.68	.29	7.82	.42	9.80	.46	9.58
8	weight	$Y_A > Y_B = Y_C$	.28	6.33	.39	8.73	.51	10.05	.55	10.04
6	area	$Y_A > Y_B > Y_C$	.40	8.02	.48	9.88	.57	10.35	.59	10.03
Total Item Set			.09	7.40	.20	13.55	.40	17.57	.52	17.74

Item Analysis 2: Analysis of Ten Items Without the Pseudo Tasks

Step 2.1: Inspecting the Sign of the Inter-Item Covariances. A second analysis was done on the set of 10 items without the two pseudo tasks. Nine out of 45 covariances were negative. Item 4 had five negative covariances with other items, and Item 2 had four negative covariances. Because the mutual covariance of these two items was positive, this means that each negative covariance involved either Item 4 or Item 2. These two items were already suspected in the first item analysis round, based on the P values of these items in different grades.

Step 2.2: Inspecting the Scalability of Individual Items. An inspection of Table 3.2 (Second Analysis) shows that the H_i values of most items increased after removing the two pseudo tasks. Items 4 and 2 have the lowest H_i values, which confirms our doubts about the fit of Items 4 and 2 in this test. Moreover, Items 4 and 2 are the only items that deal exclusively with equality relations between the objects. Together with the numerical results obtained, this seems to provide sufficient evidence to remove these items and do a third analysis. Before doing this, we discuss a quality measure that can be used to assess a complete set of items, and then a method to estimate the IRFs.

Step 2.3: Inspecting the Scalability of the Total Item Set. So far, we have concentrated on individual items. Although the presence of low-quality items in the test has the effect of lowering total test quality, at all stages of the item analysis we would also like to know the quality of the whole test. Ultimately, this is more important than the quality of the individual items. Now we briefly discuss such a total test scalability coefficient, and then we apply it to the sets of 10 and 12 transitive reasoning items, respectively.

Theoretical Sidestep: Scalability of an Item Set. A summary of the information given by the H_is is provided by the H coefficient. This is a weighted mean of the H_is that provides evidence about the degree to which respondents can be ordered by means of the complete set of items. Under the MHM, $0 \leq H \leq 1$; thus, negative values are in conflict with the model. Furthermore, the relation between the item coefficient and the overall coefficient is that $H \geq \min(H_i)$; see Theorem 4.2 of Chapter 4. This means that if we require all H_is to be at least 0.3, then by implication also $H \geq 0.3$.

Results for Step 2.3. For the set of 10 items without the two pseudo tasks, $H = 0.20$ (with the test statistic $Z = 13.55$ indicating that this is significantly higher than 0; see Table 3.2), and for the set of 12 items, including the pseudo tasks, $H = 0.09$. These results confirm that from a practical point of view, the two item sets are not suited to order respondents, although the 10-item subset is better than the original 12-item test.

Step 2.4: Inspecting the Item Response Functions. An item scalability coefficient H_i that summarizes different aspects of individual item functioning is convenient for the researcher. However, if item functioning is not unequivocally of good quality, a single number tends to obscure important characteristics of the malfunctioning. Such characteristics are better visible when we consider an estimate of the whole IRF. In this subsection, we first discuss how to estimate the IRF and then we take a closer look at the estimated IRFs of the transitive reasoning items.

Theoretical Sidestep: Estimates of the IRFs. The almost zero H_i values of Items 4 and 2 in Item Analysis 1 and Item Analysis 2 seem to suggest that these items have very weak discrimination power. Possible causes of a low H_i value may be (a) an almost flat IRF; (b) an irregular IRF that jumps up and down across θ; or (c) a single-peaked IRF. A diagnosis of the misfit of item i thus might include the investigation of the IRF of this item.

In principle, the IRF $P_i(\theta)$ could be estimated from the data by first estimating θ and then calculating the fraction of persons with a particular estimated θ who have a score of 1 on item i. Across the estimated θ values this would yield a discrete estimate of the IRF. In a nonparametric context, however, numerical estimates of θ are not available, and a next-best solution is to replace θ with the total score on the other $k - 1$ items without item i. This total score is denoted $R_{(i)}$ and is referred to as the rest score. It is formally defined as

$$R_{(i)} = \sum_{j=1; j \neq i}^{k} X_j. \tag{3.1}$$

Next, conditional probabilities of obtaining a score 1 as a function of $R_{(i)}$ are estimated. Such probabilities are denoted $P_i[R_{(i)}]$ and can be estimated as sample fractions. A theoretical result says that under the MHM, for a fixed item the probabilities $P_i[R_{(i)}]$ are nondecreasing in $R_{(i)}$. This is the property of manifest monotonicity, which means that the regression of the score on item i on the rest score is nondecreasing. Throughout, this regression will be referred to as item-rest regression.

Results for Step 2.4. The item-rest regressions of all 10 remaining items are displayed in Table 3.3. To obtain the sample fractions, groups of respondents were formed on the basis of $R_{(i)}$ and then adjacent rest score groups were joined if they contained fewer than 20 respondents until a joined group resulted that had at least 20 respondents. The minimum group size of 20 was chosen because few respondents had very low rest scores, and as a result, sample fractions would estimate probabilities $P_i[R_{(i)}]$ very inaccurately. This would result in little power to detect reversals in the expected nondecreasing regression in the lower groups. Our choice forced the lowest rest score groups to be joined and left the higher, much larger groups intact. The fractions were calculated in the groups after joining. As an example, Table 3.4 contains the observed frequencies necessary to produce the fractions for Item 9 in the first column of Table 3.3. Exercise 3.3 asks the reader to perform the calculations.

The item-rest regression of Item 4 had an irregular pattern, although none of the reversals of the expected ordering was significant at a 5% level (using a test based on an accurate normal approximation of the hypergeometric distribution; see Molenaar & Sijtsma, 2000). This result provides evidence that the IRF is almost horizontal and that the item has no relation to the latent trait as measured by the other items. The item-rest regression of Item 2 shows only one small reversal, from 0.76 to 0.75, but it increases slowly across $R_{(i)}$, thus suggesting weak discrimination and a weak relation to the latent trait.

Table 3.3   Item-Rest Regressions of Ten Items; Second Analysis.

	Item Number									
Group	9	10	4	5	2	7	3	1	8	6
0–4			.75	.60	.76	.40	.43	.67	.73	.71
0–5	.13	.26								
5			.80	.71	.75	.71	.76	.93	.91	.95
6	.12	.36	.73	.75	.76	.76	.83	.92	.97	.99
7	.28	.48	.80	.82	.80	.88	.93	.97	1.00	1.00
8	.29	.55	.79	.82	.83	.93	.96	.98	.99	1.00
9	.45	.78	.80	.93	.89	.98	.98	.98	1.00	1.00

Table 3.4   Number of Correct Answers on Item 9, and Total Number of Observations Within Rest Score Groups Defined by $R_{(9)}$.

	$R_{(9)}$									
	0	1	2	3	4	5	6	7	8	9
$X_9 = 1$	0	0	1	1	1	1	6	27	40	51
Total	0	1	4	3	6	17	48	97	136	113

None of the reversals of the expected ordering for the other items was significant, so this provides no evidence against the monotonicity assumption of the MHM. Note that the easy items, such as Items 1, 8, and 6, have item-rest regressions that start at a high probability level and do not distinguish between the higher rest score groups.

Item Analysis 3: Also Deleting Items 2 and 4

The third item analysis concerned eight items, omitting the two pseudo tasks and Items 4 and 2. No negative covariances were found between the remaining $\binom{8}{2} = 28$ item pairs. Seven items had H_i values higher than 0.3 (Table 3.2, Third Analysis). The overall H was 0.40. None of the item-rest regressions showed significant reversals.

For Item 5, the item scalability coefficient is $H_5 = 0.20$, and according to the rule of thumb for admission to the test, this is too low. At least two courses of action are possible. The first is to retain Item 5, because in a short test, each item that has at least some positive discrimination power (see also the increasing item-rest regression of Item 5 in Table 3.3) may contribute to the reliability of the total score. The second is to study Item 5 further in an effort to learn more about the measurement of transitive reasoning. We follow the second option.

Item Analysis 4: Deleting Item 5

It may be noted that Item 5 is the only item in which only smaller-than ($<$) relations between objects are studied rather than larger-than ($>$) relations. This makes a difference in the conversation of the experimenter and the child: For Item 5, in conversation, the relations were indicated as "lighter" rather than "heavier" (Item 3), "longer" (Item 1), and "larger" (Item 6). Items 9 and 10 contain one smaller-than relation in addition to

Table 3.5   Frequency Distribution of Total Score X_+ Based on Seven Items; Mean, Standard Deviation, Skewness, and Kurtosis; and Reliability of X_+.

X_+	0	1	2	3	4	5	6	7
Frequency	3	4	7	17	40	130	138	86

Mean	5.4	Skewness	−1.2
Standard Deviation	1.3	Kurtosis	2.4
Reliability	0.68		

two equalities. Usually, tasks having mixtures of relations are distinguished from tasks that have only one type (e.g., Verweij, Sijtsma, & Koops, 1999). It may be hypothesized that this format ($<$ for "lighter") was unexpected for some respondents, thus creating confusion that led them to solve the problem as if a heavier-than relation had been presented.

Leaving out Item 5 and reanalyzing the data for the remaining seven items led to an increase of all H_is (Table 3.2, Fourth Analysis). The overall H increased from 0.40 (Third Analysis) to 0.52. These results again show a major improvement of the psychometric quality of the test. It seems safe to conclude that Item 5 did not fit in with the other items.

Scale Score Statistics

Finally, we provide some psychometric properties for the set of seven items obtained in the Fourth Analysis. Table 3.5 shows that most respondents have a high X_+ value. This is also clear from the mean of the distribution. The skewness is slightly negative. The reliability of X_+ is equal to 0.68. This estimate is based on a method that assumes that the IRFs are nonintersecting, a property that is not investigated here but will be studied extensively in Chapter 6.

Measures Against Chance Capitalization

The previous analyses have made clear that item analysis under an IRT model typically involves a large number of steps and a large number of decisions per step. The higher these numbers, the more realistic the threat of chance capitalization; that is, one views as systematic and important certain aspects or differences that are caused purely by random variation. In particular, when sample size is small, say, no more than 500, chance capitalization is a realistic threat and may distort the results of an item analysis. For these reasons, we believe that a textbook on test construction like the present one should warn against chance capitalization and provide solutions. Here are a few.

One way to assess the effects of repeatedly relying on the results of the previous step is to split the sample into two halves before the item analysis begins, and to perform the step-by-step analysis on one half and check the results on the other half. This cross-validation is important in any item analysis performed under any model. When we did our split, all main conclusions for the four analyses remained unaltered for each half sample (the total H differed by no more than 0.02 between halves, and the item H_i by no more than 0.10). When the reader has access to MSP, such a split can be done conveniently by means of a grouping variable distinguishing the halves.

Another possibility is to collect data in another sample from the same population and repeat the analyses for these new data. For this purpose, we used data that were collected 4 months after collecting the data set analyzed here. This second data set was collected in one half of the *same* group as well as in a sample of 209 *newly* drawn respondents to assess whether the test could be used in longitudinal research. We used the new sample of 209 respondents and repeated most of the analyses reported in this chapter. In consecutive analysis steps, first the two pseudo tasks, then Items 2 and 4 and, finally, Item 5 appeared as the deviant items. This resulted in the same final seven-item scale as found in this chapter: For this scale, $H = 0.63$ (compare $H = 0.52$ for the original group of respondents) and at the item level $0.44 \leq H_i \leq 0.79$ (compare $0.46 \leq H_i \leq 0.59$ for the original respondent group; also see Table 3.2). In general, based on this new sample, we drew the same conclusions as reported in this chapter for the first data set.

If only one sample is available, it is possible to protect against chance capitalization by adapting the significance level of statistical tests to the number of decisions made on the basis of these tests. This adaptation controls the Type I error rate. Note, however, that chance capitalization occurs due to consecutive decision making on the basis of a sample and that it will occur irrespective of whether decisions are based on formal statistical testing or not. Item analysis typically involves both statistical testing and decision making based on descrip-

tive statistics and also on the content of the items. Adapting the significance level thus leads to control only over Type I errors in statistical
testing, but not over chance capitalization due to other decisions, for
example, based on descriptive statistics. It may be noted that the use of
item content may even protect the researcher from taking the numerical
sample results too seriously.

Another protection is to perform additional research aimed at specific
questions related to the test construction. For example, in a controlled
experiment, Verweij (1994, pp. 65–69) studied two versions of Item 5,
here denoted Item 5 and Item 5^*. The original Item 5 had a format that
was the reverse of that of the other items (see Table 3.1), whereas Item 5^*
had the same format as the other items. Including Item 5^* instead of Item
5 with the other seven items led to $H_{5*} = 0.37$ (cf. Table 3.2, where
results for H_5 [reversed format] are given), a value much more like the H_is
of the other items found in the experiment. This result, and several results
from other research (Verweij et al., 1999), finally led to a new, balanced
nine-item test version (due to the complete crossing of two task characteristics, the number of objects in a task [3, 4, and 5] and the property to be
assessed [length, size, and weight]), with $H = 0.75$, H_is between 0.62 and
0.84, and item-rest regressions that were nondecreasing.

Discussion

Without further external evidence of their validity, and on the basis of
our results alone, we assume that the seven remaining items together
constitute a short test for the measurement of transitive reasoning about
inequality relations and mixtures of inequalities and equalities. Given
that the MHM holds for this item set, the 86 children (Table 3.5) with
$X_+ = 7$, for example, have a mean θ that is at least as high as the mean
θ of the 138 children with $X_+ = 6$; this latter group has a mean θ that is
at least as high as that of the 130 children with $X_+ = 5$; and so on. This
ordering result follows the stochastic ordering property of the MHM
and was formalized in Equation 2.10. The estimated reliability equaled
0.68 (the method of reliability estimation is discussed in Chapter 6).
Because diagnostic tests usually are required to have a reliability of at
least 0.9 (Nunnally, 1978, p. 246), the transitive reasoning test constructed may not be suited to this purpose. A test suitable for individual
diagnosis of transitive reasoning ability should have more items to
increase reliability. Moreover, for our population, the difficulty level

of the test should be higher than that of the seven-item test found here (Table 3.5).

Additional Reading

The transitive reasoning data used in this chapter came from research by Verweij et al. (1996). Other examples of item analysis using the MHM can be found in Appendix 1. Properties of the MHM relevant to model-data fit research are discussed by Mokken and Lewis (1982) and Sijtsma (1998). The item H_i coefficient and the overall H coefficient are discussed by Mokken (1971) and Mokken and Lewis (1982). The estimation of the IRF is discussed in a mathematical paper by Junker and Sijtsma (2000). All methods used in this chapter are discussed and applied to real data in the MSP manual (Molenaar & Sijtsma, 2000).

Exercises for Chapter 3

3.1. a. Which statistics and methods can be used to assess the quality of an item for inclusion in a scale under an MHM analysis?
   b. Give the interpretation of each of the statistics mentioned under (a).

3.2. a. Which scalability coefficients have been distinguished?
   b. How do they differ?
   c. Given the H_is in Table 3.2, Second through Fourth Analysis, what can be said about the minimum value of H based on a knowledge of only the H_is?

3.3. a. Use the data in Table 3.4 to calculate the fractions in the first column of Table 3.3 that concern the item-rest regression of Item 9.
   b. For Item 9, which groups would have to be joined if the minimum group size were 10? And which groups if the minimum group size were 50?
   c. For Exercise b, calculate the two corresponding fractions.
   d. For minimum group sizes of 10 and 50 explain, relative to a minimum group size of 20 (Table 3.3), what the possible weaknesses and merits of each choice would be.
   e. Calculate all fractions for minimum group size 1, and interpret the instability of the first five values obtained.

3.4. Which strategies are recommended to protect against chance capitalization in a step-by-step item analysis?

Answers to Exercises for Chapter 3

3.1. a. P_i-value, H_i-value, item-rest regression

   b. P_i: proportion of 1 scores on item i, or mean item score; H_i: indicates how well item i fits together with the other items for the purpose of person ordering on a latent trait; item-rest regression: conditional probability of a score of 1 as a function of the rest score.

3.2. a. H_i and H

   b. H_i gives the scalability of item i with respect to the other items, and H, which is a weighted mean of the H_is, gives the scalability of the total item set.

   c. Because $H \geq \min(H_i)$, H is at least 0.04, 0.20, and 0.46, respectively.

3.3. a. The first fraction of the first column of Table 3.3 (Item 9) is calculated from $(0 + 0 + 1 + 1 + 1 + 1)/(0 + 1 + 4 + 3 + 6 + 17) \approx 0.13$; the second fraction is $6/48 = 0.125$ (rounded to 0.12); and so on.

   b. Minimum group size 10: $0 \leq R_{(9)} \leq 4$; minimum group size 50: $0 \leq R_{(9)} \leq 6$.

   c. Size 10: $3/14 \approx 0.21$; Size 50: $10/79 \approx 0.13$.

   d. Size 10: weakness is less power to detect deviations, merit is more points of item-rest regression available. Size 50: weakness is fewer points of item-rest regression available, merit is more power.

   e. The fractions are: not defined, 0.00, 0.25, 0.33, 0.17, 0.06, 0.12, 0.28, 0.29, 0.45. Given the small group sizes, the first five fractions are very inaccurate estimates of the population proportions.

3.4. a. Split sample into two halves, perform item analysis on one half, and check results on the other half; (b) check results on new sample; (c) adapt significance level of statistical tests; and (d) resolve questions raised during item analysis by follow-up (quasi-)experimental research.

4

The Monotone Homogeneity Model: Scalability Coefficients

In this chapter we explain

- how to estimate the ordering of items and persons
- how to calculate the scalability coefficients H (for each item pair, for each item, and for the total scale)
- to what extent these coefficients provide information on the presence or absence of a useful scale satisfying the MHM

It is important to formulate these procedures and properties in a general and precise way. This is a prerequisite for fully understanding them regardless of specific examples or applications. Therefore, the presentation in this chapter is a bit more formal and less narrative. We recommend that readers who are not familiar with formulas move more slowly and use the examples and exercises to gain more insight.

Ordering of Items and Persons

In Chapters 2 and 3, we ordered items by their proportion-correct P_i. Now we consider these orderings in more detail. In practice, they are based on results for a sample of persons; it is legitimate to ask whether they can be generalized to the population of all respondents to which the test or scale could be administered. In the section on nonintersection of IRFs at the end of Chapter 2, we saw that items can be ordered by the proportion-correct in the whole population. If we have a random sample of n respondents, for each item the fraction-correct in the sample is an unbiased and consistent estimator of the population proportion. In repeated samples from the same population, the number of persons with item i correct has a binomial distribution. Via this binomial distribution (or its normal approximation for large n and proportions not too close to 0 or 1), one may obtain confidence intervals for the population values of the item popularities.

There is some risk of a reverse ordering for an item pair in the sample compared to the population. Suppose that in the population of all respondents, item i has a proportion correct of 0.50 and item j has a proportion-correct of 0.51. If we take a random sample of size 10, say, then it could easily happen that in the sample $P_i > P_j$. For example, 6 out of 10 persons may score positively on item i and only 5 out of 10 on item j. The researcher would then erroneously conclude that item i is easier than item j. Later in this chapter, this risk of reverse ordering in the sample compared to the population will be further discussed. We will see that it is unwise to use samples of only 10, but that the risk is small when several hundreds of respondents are used.

As was announced in the Models and Data section of Chapter 2, the total score X_+ is used to estimate the ordering of the respondents by their latent trait value θ, which is, of course, not observable. Recall that Equations 2.9 to 2.12 show that apart from random fluctuations, there is indeed agreement in the MHM between ordering by total score and ordering by latent trait value. In most applications of IRT, however, the number of items is much lower than the number of subjects, and the person ordering is less reliably reflected in the data than the item ordering. This issue will be further discussed in Chapter 6 in the context of reliability of the total score X_+.

Nonnegative Covariances

If the MHM holds for a test consisting of k dichotomous items, we know, in the notation already used in Chapter 2, that the IRFs,

$$P_i(\theta) = P(X_i = 1|\theta), \text{ for } i = 1, \ldots, k, \tag{4.1}$$

are nondecreasing in θ. This, of course, is an assumption that cannot be directly verified because θ is latent, not observable. We now present an important observable consequence of the MHM; its proof can be found in Appendix 4.

THEOREM 4.1: If the MHM holds, that is, if the assumptions of unidimensionality, local independence, and monotonicity of Chapter 2 are fulfilled, then the covariances between all item pairs are nonnegative.

This important theorem will be heavily used throughout this book. For two reasons, however, it does not solve all problems involved in checking the monotonicity of IRFs. First, the condition of nonnegative covariances for item pairs is necessary but not sufficient for nondecreasing IRFs: It is easy to construct counterexamples in which such a curve decreases on a small interval of θ, whose negative contribution to the covariance with any other item is compensated for by a stronger positive contribution of other θ intervals. Second, sampling error may cause a covariance estimate to be negative even though the population covariance is positive, and vice versa. A significance test based on a normal approximation is implemented in the MSP software.

Scalability of an Item Pair: Illustrations

In Chapter 3, we met the use of a coefficient H_i expressing, loosely speaking, the extent to which item i fits together with the other $k - 1$ items of a scale, for the purpose of ordering persons on the intended latent trait. At the end of this chapter, we discuss in more detail what we mean by this concept of "scalability." First, however, we explain how to calculate these coefficients. The scalability coefficient is best understood by first considering the scalability of an item pair. After that, the scalability of each item can be reduced to the scalability of all item pairs in which it occurs, and the scalability of the whole test or scale can be reduced to the scalability of all pairs of items. Here, two numerical examples are given; the general definition of the coefficient per item pair in the population follows in the next section.

In the example presented in Chapter 3, the four possible outcomes for Items 9 and 6 can be arranged in a 2×2 table of observed frequencies (Table 4.1). The sample covariance is $0.2988 - 0.3012 \times 0.9741 = 0.005442$, and the sample correlation is 0.0747: Both outcomes are positive but very small. Given the difference in item popularity, however, a high correlation is simply impossible. The best possible result would hold if not 1 but 0 respondents had scored $X_9 = 1$ and $X_6 = 0$; see Table 4.2. This fictitious table has a sample covariance of 0.007795 and a sample correlation of 0.107012.

Table 4.1   Observed Joint Frequencies of Items 6 and 9.

	Frequencies				Sample Fractions		
	$X_9 = 0$	$X_9 = 1$	Total		$X_9 = 0$	$X_9 = 1$	Total
$X_6 = 0$	10	1	11	$X_6 = 0$	0.0235	0.0024	0.0259
$X_6 = 1$	287	127	414	$X_6 = 1$	0.6753	0.2988	0.9741
Total	297	128	425		0.6988	0.3012	1.0000

Table 4.2   Joint Frequencies of Items 6 and 9 (Maximum Correlation).

	Frequencies				Sample Fractions		
	$X_9 = 0$	$X_9 = 1$	Total		$X_9 = 0$	$X_9 = 1$	Total
$X_6 = 0$	11	0	11	$X_6 = 0$	0.0259	0.0000	0.0259
$X_6 = 1$	286	128	414	$X_6 = 1$	0.6729	0.3012	0.9741
Total	297	128	425		0.6988	0.3012	1.0000

The scalability coefficient of the item pair $(9, 6)$ is now calculated as the ratio of the two outcomes; because the marginals (row and column totals) are kept fixed, the standard deviations of each item are the same in both tables and we obtain

$$H_{9,6} = \frac{Cov}{Cov_{max}} = \frac{0.005442}{0.007795} = \frac{r}{r_{max}}$$

$$= \frac{0.07471}{0.10701} = 0.698.$$

(4.2)

Note that this expression would have been written as φ/φ_{max} in the older psychometric literature where a product-moment correlation between two dichotomous variables was denoted by φ; see also Equation 2.21 in Exercise 2.1 of Chapter 2.

This procedure for finding the pairwise H coefficient for two dichotomous items can be summarized in seven steps:

Step 1. Obtain the 2×2 table for the item pair

Step 2. Obtain the correlation r between the item scores

Step 3.　Find the error cell (easier item incorrect, more difficult item correct)

Step 4.　Change its observed frequency to 0

Step 5.　Change the other three frequencies accordingly, keeping the marginal entries of the table fixed

Step 6.　Obtain the correlation r_{max} in the new 2 × 2 table

Step 7.　Find $H = r/r_{max}$

The same answer can be found after replacing the correlation by the covariance in Steps 2, 6 and 7. This may be somewhat easier in hand calculation and somewhat more complicated with standard statistical software. Before we illustrate the seven steps for another item pair (1, 8) from the same transitive reasoning data (Chapter 3), we first consider an alternative way to obtain H.

In the perfect Guttman scalogram, as the reader possibly knows, it never occurs that a subject scores 0 on a more popular (i.e., easier) item and 1 on a less popular (i.e., more difficult) item. These reversals are called "Guttman errors." In our example (Table 4.1), the cross tabulation of Items 9 and 6 shows that $F_{6,9} = 1$: Only 1 respondent has Item 9 correct and the much easier Item 6 incorrect, which is a Guttman error. For the given marginals, the minimum of $F_{6,9}$ would be 0 (as in Table 4.2) and the maximum would be 11. Now consider as a kind of boundary case the situation in which both item scores are statistically independent, not given θ (local independence) but in the whole population. We would then expect a frequency of $E_{6,9} = 11 \times 128/425 = 3.3129$ in the "error cell" defined by $X_9 = 1$ and $X_6 = 0$. In order to obtain a coefficient that equals 1 for perfect scalability and 0 for statistical independence, divide the two error counts and subtract from 1:

$$H_{6,9} = 1 - \frac{F_{6,9}}{E_{6,9}} = 1 - \frac{1}{3.3129} = 0.698. \qquad (4.3)$$

We show below that it is no coincidence that the definitions in Equations 4.2 and 4.3 lead to the same value. This new calculation method can be summarized in five steps:

Step 1F.　Obtain the 2 × 2 table for the item pair

Step 2F.　Find the error cell (easier item incorrect, more difficult item correct)

Step 3F.　Find its observed frequency F

Step 4F. Find its expected frequency E under the null model of independence keeping the observed marginals fixed

Step 5F. Obtain $H = 1 - F/E$

Both procedures, using r and using F, are now applied to another item pair. From the data set used in Chapter 3, we choose a pair with almost equal item popularities. Item 8 has a sample fraction of $411/425 = .9671$ correct answers, and Item 1 has a fraction $400/425 = .9412$. In Step 1, the 2×2 table is found (Table 4.3). Step 2 calculates its sample correlation to be 0.2340. The error cell has Item 1 correct and the slightly easier Item 8 incorrect (Step 3). We change its frequency, which equals 9 in Table 4.3, to zero (Step 4) and in Step 5 obtain the other three frequencies, keeping the marginal entries fixed (Table 4.4). That table has $r_{max} = 0.7382$ (Step 6). Finally, Step 7 finds $H_{1,8} = r/r_{max} = 0.317$.

For the second method, the calculation using F, Step 1F finds the 2×2 table (Table 4.3) and Step 2F its error cell, as before. It contains $F = 9$ observations (Step 3F). In Step 4F, we keep the marginals 400 (for $X_1 = 1$) and 14 (for $X_8 = 0$) fixed and obtain an expected frequency under statistical independence of $E = 400 \times 14/425 = 13.1765$ for this error cell. Step 5F gives $H_{1,8} = 1 - 9/13.1765 = 0.317$.

It is interesting to compare the results for the two item pairs (6, 9) and (1, 8). The latter has a correlation coefficient $r = 0.2340$. This is about three times as high as for (6, 9) where $r = 0.0747$. The maximum possible

Table 4.3   Observed Joint Frequencies of Items 1 and 8.

	$X_8 = 0$	$X_8 = 1$	Total
$X_1 = 0$	5	20	25
$X_1 = 1$	9	391	400
Total	14	411	425

$$r = 0.2340$$

Table 4.4   Joint Frequencies of Items 1 and 8 (Maximum Correlation).

	$X_8 = 0$	$X_8 = 1$	Total		$X_8 = 0$	$X_8 = 1$	Total
$X_1 = 0$	—	—	25	$X_1 = 0$	14	11	25
$X_1 = 1$	0	—	400	$X_1 = 1$	0	400	400
Total	14	411	425		14	411	425

$$r = 0.7382$$

correlation given the marginals, however, is $r_{max} = 0.7382$ for (1, 8) and only $r_{max} = 0.1070$ for (6, 9). The two examples show that a low correlation between two dichotomously scored items may be due mostly to a dramatic difference in item popularity. Traditional methods such as factor analysis or cluster analysis based on correlation coefficients would probably not include items 6 and 9 in the same scale. Our nonparametric item selection procedure (discussed in detail in Chapter 5), however, is based on the pairwise H coefficient that takes the observed difference in item popularity into account. This is expressed, in our example, by a pairwise H of 0.698, which is more than twice as high as 0.317: The item pair (6, 9) fits together better, in the sense of H, than the item pair (1, 8).

Scalability of an Item Pair: Theory

In the general case, consider a pair of items, say, item i and item j; let i denote the more difficult item. Then $P_i < P_j$ in the notation of Equation 2.15. Table 4.5 gives the 2×2 table of probabilities (left) and the table producing the maximum correlation given the marginals (right).

Further, let P_{ij} denote the probability of correct (1) scores on both items i and j. The top right cell of Table 4.5 indicates the probability of $X_i = 1$ and $X_j = 0$, which is zero if and only if $P_{ij} = P_i$ (no Guttman errors). We now define (first equality of Equation 4.4) the H_{ij} coefficient as

$$H_{ij} = \frac{Cov(X_i, X_j)}{Cov_{max}(X_i, X_j)} = \frac{P_{ij} - P_i P_j}{P_i - P_i P_j} = 1 - \frac{P_i - P_{ij}}{P_i(1 - P_j)}, \quad (4.4)$$

and it follows from simple algebra that the covariance divided by its maximum, given the marginals, is exactly equal to 1 minus the ratio of the actual probability of a Guttman error (value $P_i - P_{ij}$, sample estimate F_{ij}/n)

Table 4.5 Model Probabilities for the General Case and for Maximum Correlation.

	Model Probabilities			Maximum Correlation		
	$X_i = 0$	$X_i = 1$	*Total*		$X_i = 0$	$X_i = 1$	*Total*
$X_j = 0$	$1 - P_i - P_j + P_{ij}$	$P_i - P_{ij}$	$1 - P_j$	$X_j = 0$	$1 - P_j$	0	$1 - P_j$
$X_j = 1$	$P_j - P_{ij}$	P_{ij}	P_j	$X_j = 1$	$P_j - P_i$	P_i	P_j
Total	$1 - P_i$	P_i	1	Total	$1 - P_i$	P_i	1

and the probability expected under the null hypothesis of statistical independence between the scores of items i and j, which has the value $P_i(1 - P_j)$ and sample estimate E_{ij}/n; see the final part of Equation 4.4.

The population value of the pairwise H coefficient of two items has thus been defined in close analogy to its sample counterpart. It should be noted that the sample H conditions on the observed marginal distributions (i.e., the sample fractions, P, of both items), in determining which cell in the 2×2 table is the error cell, in calculating the maximum possible covariance or correlation, and in calculating the expected number of Guttman errors under independence. Such a conditioning on sufficient statistics for nuisance parameters is often useful in statistical methodology and follows the same reasoning that is used for defining the chi-square test statistic for general frequency count tables, where the marginals are fixed for calculating the expected cell frequency counts under the null model.

Of course, repeated samples from the same population will often produce somewhat different sample marginals. If two items have almost the same popularity in the population, and sample size is moderate or small, then in some samples, the popularity order of the two items may be reversed, after which the (0, 1) cell rather than the (1, 0) cell is viewed as the error cell by a researcher who sees only this sample. This is potentially a source of wrong inferences. By enumerating all possible 2×2 tables, however, it can be shown that the negative effects of treating the item marginals as fixed are usually minor. That somewhat technical analysis, not presented in this book, shows that the effect of conditioning on incorrectly ordered marginals is almost always negligible for sample sizes $n > 100$ when each pair of item popularities differs by more than 0.02. This bound becomes $n > 200$ when incidental pairs lie more closely together and $n > 400$ with many item pairs at such small distances.

Although these sample size bounds may be conservative, it is obvious that a data set with, for example, 40 items and 40 persons does not permit accurate estimation of the item ordering, nor of the item popularities, the H coefficients, or other statistics discussed in later chapters. Consequently, decisions concerning item ordering, item fit, and model fit then become rather unstable.

Scalability of an Item and of the Test

For the definition of the scalability coefficient H_i of item i, recall that in Equation 3.1, we defined the rest score $R_{(i)}$ as the sum score across the $k - 1$

other items. In this subsection, we assume that the items are numbered from lowest to highest item popularity, as in Equation 2.19, because this simplifies the notation. This means that for $j > i$, the "error cell" of the item pair (i, j) is $X_i = 1$ and $X_j = 0$, but for $j < i$, it is $X_j = 1$ and $X_i = 0$. Analogous to the pairwise H, we now introduce the item scalability coefficient,

$$H_i = \frac{\text{Cov}(X_i, R_{(i)})}{\text{Cov}_{\max}(X_i, R_{(i)})}$$

$$= \frac{\sum_{j \neq i}(P_{ij} - P_i P_j)}{\sum_{j > i}(P_i - P_i P_j) + \sum_{j < i}(P_j - P_i P_j)} \qquad (4.5)$$

$$= 1 - \frac{\sum_{j \neq i}(P_i - P_{ij})}{\sum_{j > i}P_i(1 - P_j) + \sum_{j < i}P_j(1 - P_i)}.$$

This population value can be estimated by

$$\hat{H}_i = 1 - \frac{\sum_{j \neq i} F_{ij}}{\sum_{j \neq i} E_{ij}}. \qquad (4.6)$$

That is, the sum frequency of all Guttman errors in which item i is involved must be divided by its expected value assuming statistical independence between item i and all other items, and the result must be subtracted from 1.

As a final step, the scalability of all k items jointly is defined by

$$H = \frac{\sum_i \text{Cov}(X_i, R_{(i)})}{\sum_i \text{Cov}_{\max}(X_i, R_{(i)})}$$

$$= \frac{\displaystyle\sum_{i}\sum_{j \neq i}(P_{ij} - P_i P_j)}{\displaystyle\sum_{i}\sum_{j>i}(P_i - P_i P_j) + \sum_{i}\sum_{j<i}(P_j - P_i P_j)} \qquad (4.7)$$

$$= 1 - \frac{\displaystyle\sum_{i}\sum_{j \neq i}(P_i - P_{ij})}{\displaystyle\sum_{i}\sum_{j>i}P_i(1 - P_j) + \sum_{i}\sum_{j<i}P_j(1 - P_i)}.$$

This scalability coefficient is estimated in the sample from the ratio of the sum of all observed Guttman errors (now across all item pairs) to the corresponding sum expected under independence of all item pairs given the item marginals, and the result is subtracted from 1.

Properties of the Scalability Coefficients

In this section, we present further properties and interpretations of the scalability coefficients. First, we prove two important theorems.

THEOREM 4.2: For all indices (i, j),

$$\min_{j} H_{ij} \leq H_i \leq \max_{j} H_{ij}; \qquad (4.8)$$

$$\min_{i} H_i \leq H \leq \max_{i} H_i; \qquad (4.9)$$

and

$$\min_{i,j} H_{ij} \leq H \leq \max_{i,j} H_{ij}. \qquad (4.10)$$

PROOF. It follows from Equation 4.5 that H_i can be written as a weighted sum of the H_{ij}s, with weights proportional to the (positive) denominators. Similarly, it follows from Equation 4.7 that H is a weighted sum of both the H_is and of the H_{ij}s. This establishes that the three inequalities follow from the definitions of the H coefficients; thus, their validity is not restricted to item sets for which the MHM (or the DMM) holds.

THEOREM 4.3: If the MHM holds, then the population H values of item pairs, items, and the total test satisfy

$$0 \le H_{ij} \le 1, \text{ for all } i, j = 1, \ldots, k, i \ne j; \qquad (4.11)$$

$$0 \le H_i \le 1, \text{ for all } i = 1, \ldots, k; \qquad (4.12)$$

and

$$0 \le H \le 1, \qquad (4.13)$$

respectively.

PROOF. Earlier in this section, we saw that the MHM implies that the covariance of each item pair is nonnegative in the population. The bounds of 0 and 1 thus hold for the pairwise H_{ij}, which is $\text{Cov}(X_i, X_j)/\text{Cov}_{max}(X_i, X_j)$. Because the item H and total scale H are positively weighted sums of H values for item pairs, they are also bounded by 0 and 1, which completes the proof.

The boundary case $H = 1$ means "no Guttman errors." When $H = 0$ for an item pair, it means "no correlation"; for an item or the total test it means that the corresponding average of the covariances is zero. A theoretically interesting situation occurs when all persons have the same latent trait value; see Exercise 4.7.

Negative H values can occur for two reasons. The first is sampling fluctuation: For a small n, a positive population value may have a negative estimate. Mokken (1971, pp. 160–164) presented the exact variance of a sample H given the marginals, and used it to obtain a Z value that is approximately normally distributed under the hypothesis that $H = 0$ in the population. The formulas are not discussed here; they are implemented in the MSP software mentioned earlier.

The second reason one obtains a negative H value is violation of the MHM. If this model fails to hold, the pairwise H_{ij} can be negative, and depending on the P_i values, even smaller than -1. The actual minimum, which corresponds to the maximum possible frequency in the error cell given the marginals, depends in a complicated way on the values of those marginals, as is illustrated in the exercises. The minimum under the MHM is zero (see Theorem 4.3), apart from sampling fluctuations. If the items of a scale were carefully selected, negative H values will be rare, and items for which they occur may be eliminated or changed; see Chapters 3 and 5.

Discussion of the Scalability Concept

In all applications of the MHM, the H coefficient plays a central role. Originally, it was primarily viewed as a measure to express the extent to which a given data set approximates the ideal of a perfect Guttman scalogram, for which $H = 1$.

More by coincidence than by principle, the same H coefficient is a useful quick-and-dirty measure for a more practical aspect of measurement quality: whether the items have enough in common for the data to be explained by one underlying latent trait in an MHM in such a way that ordering the subjects by total score is meaningful. For the ordering to be reliable as well, positive H_{ij} values for each item pair are not enough. What is needed in addition is a rule of thumb that plays a role similar to the customary requirements for the reliability coefficient $\rho_{XX'}$ in classical test theory. Many psychologists would say that $\rho_{XX'}$ must be at least 0.9 for important decisions about a person, say, with respect to admission or selection, and at least 0.7 or 0.6 for valid inferences about groups of persons.

Similarly, Mokken (1971, p. 185) proposed that a scale is useful only if $H_i \geq c$ for all items, where the lowerbound c must be specified by the user but should be at least 0.3. A scale is considered weak when $0.3 \leq H < 0.4$ for the total item set, medium when $0.4 \leq H < 0.5$, and strong when $H \geq 0.5$. According to Theorem 4.2, the choice of the lowerbound c reflects how much cohesion between items is required (also, see Chapter 5). If most H_i values were to lie between 0 and 0.3, as was the case in the First Analysis presented in Table 3.2, then the items would not have enough in common to trust the ordering of persons by total score to accurately reflect an ordering on a meaningful unidimensional latent trait.

It sometimes occurs that almost all the variance in the observed scores of the respondents is due to measurement error, not to a true variation in latent trait values. This may even happen when the scale is of good quality but the group being considered is homogeneous. More often it means that most θ values fall in a region where the IRFs are almost horizontal. In both cases, it would be very misleading to conclude that measurements on such a scale really do discriminate between respondents.

Note that the numerical value of the various H coefficients is strongly influenced by the probability distribution of the latent trait values of the respondents, but the basic model assumptions of unidimensionality, local independence, and monotonicity are not. This reflects the fact that a set of items whose IRFs are nondecreasing, possibly also nonintersecting, can be suitable for measurement in some populations but not in others. The additional requirements, like $H_i > 0.3$, that are not part of the formal MHM assumptions, serve to prevent us from using the model in a situation where it might be formally correct but substantively almost meaningless.

Additional Reading

After its invention (Loevinger, 1947, 1948), the H coefficient was of modest importance until Mokken (1971) used it for scale construction and scale evaluation. He argued from its $(1 - F/E)$ definition that it is a good yardstick for measuring the extent to which observed data approach the Guttman scalogram, and that it fulfills this role better than some alternative indices of reproducibility because it is standardized with respect to numbers of items and persons and it assumes the meaningful values of 1 when there are no Guttman violations and of 0 in case of statistical independence.

The interested reader may consult Jansen (1981, 1982a, 1982b), Jansen, Roskam, and Van den Wollenberg (1982, 1984), Mokken, Lewis, and Sijtsma (1986), Molenaar (1982a, 1982b), Roskam, Van den Wollenberg, and Jansen (1986), and Sijtsma (1984, 1986) for a discussion on the usefulness of H in an IRT context. General introductions to the theory of the MHM can be found in Mokken and Lewis (1982), Sijtsma (1998), and Molenaar and Sijtsma (2000). Snijders (2001) defined a coefficient inspired by H in a multilevel context.

Exercises for Chapter 4

4.1. Out of 500 respondents, 450 reply positively to Item 1 and only 50 to Item 2; 46 of them reply positively to both items.
   a. Complete the 2 × 2 table as in Table 4.1.

    b. Find the number F of persons who have a Guttman error; what is their score pattern?

    c. Write the 2×2 table of expected frequencies under independence given the observed marginals.

    d. Calculate the pairwise H.

    e. Still keeping the marginals fixed, find the table for which $F = 0$ and for which $F = 1$; calculate the pairwise H_{ij}s for these tables.

4.2. When out of 20 respondents, 2 reply positively to Item 1 and 18 to Item 2, which answer pattern would imply a Guttman error? Show that the number F of such persons can assume only three possible values, write out the three tables, and calculate the pairwise H in each case.

4.3. Comparing the results of Exercises 1 and 2, what can you say about the effect on H when F increases by 1? Do you think that the same effect occurs regardless of the values of the two item popularities?

4.4. a. Give an example of two items for which $H = 1$ even though the correlation is low.

    b. Can the correlation even be zero when $H = 1$? If not, why not?

    c. Can you also give an example where the correlation is 1 but H is low? If not, why not?

4.5. When the six items of a scale have $H_i = .35, .45, .30, .40, .41,$ and $.39,$ respectively, what can you say about the H of the total scale? Can you also tell which item is the most popular? Can there be a negative H_{ij} between some item pairs? Why, or why not?

4.6. In a sample of size 100, four items have popularities of $.2, .4, .6,$ and $.8,$ respectively, and frequencies of Guttman errors equal to $F_{12} = 4, F_{13} = 4,$ $F_{14} = 0, F_{23} = 6, F_{24} = 2, F_{34} = 4$ are observed. Find the item scalabilities H_i and the overall H.

4.7. In the (unlikely) case that all respondents have the same value on the latent trait, the results of this chapter lead to an interesting special case. Derive and discuss this.

Answers to Exercises for Chapter 4

4.1. a. Enter the data and complete the table: See Table 4.6 below.

    b. $F = 4$ for $X_2 = 1$ (least popular item) and $X_1 = 0$ (most popular item).

    c. See Table 4.7 below.

    d. $E_{12} = 5$ and $H_{12} = 1 - 4/5 = .20$.

    e. See Table 4.8 below.

Table 4.6

	$X_2 = 0$	$X_2 = 1$	Total	$X_2 = 0$	$X_2 = 1$	Total
$X_1 = 0$	—	—	—	46	4	50
$X_1 = 1$	—	46	450	404	46	450
Total	—	50	500	450	50	500

Table 4.7

	$X_2 = 0$	$X_2 = 1$	Total
$X_1 = 0$	$50 \times 450/500 = 45$	$50 \times 50/500 = 5$	50
$X_1 = 1$	$450 \times 450/500 = 405$	$450 \times 50/500 = 45$	450
Total	450	50	500

Table 4.8

	$X_2 = 0$	$X_2 = 1$	Total	$X_2 = 0$	$X_2 = 1$	Total
$X_1 = 0$	50	0	50	49	1	50
$X_1 = 1$	400	50	450	401	49	450
Total	450	50	500	450	50	500

$$H_{12} = 1 - 0/5 = 1.00 \qquad H_{12} = 1 - 1/5 = 0.80$$

Table 4.9

	$X_2 = 0$	$X_2 = 1$	Total	$X_2 = 0$	$X_2 = 1$	Total	$X_2 = 0$	$X_2 = 1$	Total
$X_1 = 0$	2	16	18	1	17	18	0	18	18
$X_1 = 1$	0	2	2	1	1	2	2	0	2
Total	2	18	20	2	18	20	2	18	20

$$H_{12} = 1 - 0/0.2 = 1.0 \qquad H_{12} = 1 - 1/0.2 = -4.0 \qquad H_{12} = 1 - 2/0.2 = -9.0$$

4.2. $X_1 = 1$ and $X_2 = 0$ is the Guttman pattern. Because no table entry can be larger than its row or column marginal, the error cell frequency must be 0, 1, or 2: See Table 4.9.

4.3. When F increases by 1, H decreases by $1/E$, which is 0.2 in Exercise 1 but 5.0 in Exercise 2. This shows that one person with an anomalous answer pattern can have a dramatic influence on the pairwise H when (a) the sample is small and (b) the items differ widely in popularity.

4.4. (a) An example is Table 4.8 in the answer to Exercise 1e, where $H = 1.00$ and one finds $r = 0.11$; (b) Because $H = r/r_{max}$, the correlation cannot be zero when $H = 1$; (c) for the same reason, H cannot be lower than r because r_{max} cannot exceed 1.

4.5. According to Theorem 4.2, $.30 \leq H \leq .45$. The given values provide no indication whatsoever about the item popularities, or about pairwise H values, other than that each H_i lies between the smallest and the largest H_{ij} ($j \neq i$).

4.6. $E_{12} = 12, E_{13} = 8, E_{14} = 4, E_{23} = 16, E_{24} = 8, E_{34} = 12$; meaning that

$H_1 = 1 - (4 + 4 + 0)/(12 + 8 + 4) = 1 - 8/24 = 0.67$;

$H_2 = 1 - (4 + 6 + 2)/(12 + 16 + 8) = 1 - 12/36 = 0.67$;

$H_3 = 1 - (4 + 6 + 4)/(8 + 16 + 12) = 1 - 14/36 = 0.61$;

$H_4 = 1 - (0 + 2 + 4)/(4 + 8 + 12) = 1 - 6/24 = 0.75$;

$H = 1 - (4 + 4 + 0 + 6 + 2 + 4)/(12 + 8 + 4 + 16 + 8 + 12)$
$= 1 - 20/60 = 0.67$.

Note that the overall H is not exactly equal to the arithmetic mean of the H_i because one first sums and then divides!

4.7. If there is no variance in latent trait values, local independence coincides with global independence among all respondents. Thus, all covariances between item scores are zero (Theorem 4.1). Items can still be ordered according to popularity, but ordering of persons by their total score now reflects only measurement error, not genuine latent trait differences. H values per item pair, per item, and for the scale are all zero in the population.

5

Automated Item Selection Under the Monotone Homogeneity Model

In this chapter, we discuss an automated procedure for selecting items from a larger set into clusters that measure one latent trait each, and do so with sufficient discrimination power. This procedure uses the H coefficient as a criterion for selecting or rejecting items and may be seen as an alternative to factor analysis. It applies very well to items from tests and questionnaires that are scored polytomously or even dichotomously, which is the case considered here. Especially in the dichotomous case, factor analysis is known to produce artifactual factors that are due mostly to the difference in the popularities (P_i values) of the items, which makes a linear relationship impossible. The item selection procedure discussed here is insensitive to such artifacts.

Test and Questionnaire Construction

Before constructing the final test version, a test constructor typically starts out with a preliminary set of items whose psychometric properties are unknown or doubtful. This preliminary item set may be based on a similar version of the test that was used in another population or perhaps in another country or another culture, or the item set may be partly new or even brand-new. In a pilot study, the preliminary test version is usually presented first to a relatively small sample from the population of interest, say, consisting of at least 20 and no more than 100 respondents, in an effort to get a first impression of the quality of the items and to make necessary improvements. Given the modest sample size, item quality may be assessed by means of robust statistics such as item popularities, and by frequency distributions across response options in the case of multiple-choice items. The result may be that some items appear to be too easy or too difficult, or that a particular response option is never chosen. Such items may be replaced by other, less extreme items, and inadequate response options may be reformulated. Also, respondents' feedback concerning quality of item wording and clarity of item formulation, or a content analysis of the answers given in the case of open-ended or essay questions, may provide useful information for item improvement.

On the basis of the pilot study, a final or almost final test version is composed. This version is investigated in a larger sample, say, of at least 200 but often at least 500 respondents, or even more. Because the preliminary investigation already replaced several inadequate items by what are hoped to be better items, and because the sample size now is much larger, this second item analysis can be more systematic and statistically more profound than the first. Item analysis often is done by means of an IRT model, such as the MHM, that yields valuable information about item discrimination and the dimensionality of the total test. Items having weak discrimination power provide only modest contributions to the accurate ordering of respondents on the latent trait θ by means of the total score X_+. Such items might be removed from the test, in particular when the number of strongly discriminating items is sufficiently large to obtain an accurate person ordering. Also, a total score X_+ based on items measuring different latent traits provides misleading information about the ordering on θ, when in fact several latent traits are needed to account for the data structure. When the data are found to be multidimensional, the researcher might decide to split the test into substantively meaningful subtests and to assign total scores to each separate subtest.

The process of identifying malfunctioning items or item subsets, each of which measures a different latent trait, can be very cumbersome. Chapter 3 provides an example of an item analysis of a short test for measuring transitive reasoning in which some items were too easy and others appeared to be measuring unintended abilities. This item analysis was done stepwise, starting with the complete item set, deleting items in one step and evaluating the resulting item set in the next step. In this chapter, we present an automated item selection procedure (abbreviated AISP) that is part of the computer program MSP. This procedure may be used at any stage of the research, provided that the sample contains at least, say, 100 respondents. It results in a clustering of items with reasonable discrimination power that measure the same latent trait; low-

quality items are discarded from the analysis. This AISP greatly alleviates the complexity of a stepwise approach as discussed in Chapter 3,
especially when the item set has a complex latent trait structure where
several items measure different traits, and the quality of the items varies
a great deal.

Strategies for Item Selection

In this chapter, we use K for the initial test length rather than k, to emphasize that the final test may contain fewer items (k) than the initial test.
Researchers often start out with the complete set of K items and reject
items in steps; also see Chapter 3. This is a *top-down strategy*. If the
complete item set measures several latent traits, however, the step-by-step
removal of items may be hazardous. For example, assume that we consider
all K items simultaneously. Then, it is possible that an item, say, item i, is
an adequate measure of some latent trait that is also represented by some
of the other $K - 1$ items, but not by most items. Then, item i may have an
H_i value that is too low with respect to all $K - 1$ other items currently considered, and as a result, item i may be rejected from the analysis. This may
also happen to other items measuring the same latent trait as item i, and
this means that latent traits can be lost during a top-down item analysis.

The AISP discussed here does not suffer from such problems because
it is a *bottom-up strategy:* A small kernel of k_{sel} items, to be selected on
substantive grounds by the researcher or on formal, mathematical grounds
by the AISP, is the start set. Items are selected one by one from the
remaining $K - k_{sel}$ items and added to the already selected items as long as
in the next step an item can be found that satisfies a formal scaling criterion (using the H coefficient; see Chapters 3 and 4). If not, the AISP tries
to construct a new scale from the items not already selected into the first
scale. The end result is one or more item clusters that form a sound basis
for the construction of tests satisfying the MHM.

Definition of a Scale

The AISP selects item clusters that satisfy Mokken's (1971, p. 184) definition of a scale: A scale is a set of dichotomously scored items for
which, for a suitably chosen positive constant c,

$$\rho_{ij} > 0, \text{ for all item pairs } (i, j); \tag{5.1}$$

and

$$H_i \geq c > 0, \text{ for all items } i. \tag{5.2}$$

Equation 5.1 is equivalent to $\mathrm{Cov}(X_i, X_j) > 0$; also see Theorem 4.1, which says that covariances are nonnegative under the MHM. Also, from Equation 5.1, it follows that $H_{ij} > 0$ (Equation 4.4). This implies that, within a scale, all item pairs have positive H_{ij} coefficients. Basically, Equation 5.1 says that items belonging to the same scale should measure a common latent trait.

Equation 5.2 is a practical requirement for scale construction: Its purpose is to include only items that discriminate persons reasonably well. It is up to the researcher to decide what he or she considers reasonable. By choosing a higher lowerbound c for H_i, the AISP selects items with higher discrimination power and excludes items with lower discrimination. Negative values of H_i are in conflict with the MHM; see Theorem 4.3. Therefore, it is required that $c > 0$. Several commonly used choices for c were discussed in Chapter 4, with $c = 0.3$ as the recommended *practical* minimum lowerbound.

As was shown in Theorem 4.2, $\min(H_i) \leq H \leq \max(H_i)$; thus, Equation 5.2 implies that $H \geq c$. This result shows that the individual items together determine the quality of person ordering by means of the whole test. A higher lowerbound c means a stronger scale in the sense of a more accurate ordering of the persons on the latent trait θ by means of their total score X_+ based on all selected items (Chapter 2, stochastic ordering property, Equation 2.9).

To summarize, a scale that satisfies the requirements in Equations 5.1 and 5.2 has (a) positively correlating items or, equivalently, items for which all H_{ij}s are positive; (b) items with H_is of at least c; and (c) a total H of at least c. The interpretation of these requirements is that items in the same scale should (a) measure a common trait (Equation 5.1) with (b) reasonable discrimination power determined by lowerbound c (Equation 5.2); and (c) that the test as a whole should allow for a reasonably reliable person ordering on θ using total score X_+ (implied by Equation 5.2). The AISP, to be discussed in this chapter, selects items according to the definition of a scale in Equations 5.1 and 5.2.

The AISP takes the *raw data set* as its input and reveals the dimensionality structure of these data. Then, for each dimension separately, the AISP provides information on the psychometric properties of the items and the total score X_+ based on the selected items. One possibility is that the *result* of applying the AISP to test or questionnaire data is then used for building an item bank with known psychometric (and other) properties. Such an item bank can be used for assembling tests for special purposes, typical of educational measurement. However, most applications of the AISP are in the context of constructing tests and questionnaires from raw data, not from an item bank filled with items for which all relevant properties are already known, and this makes the procedure rather different from item selection from calibrated item banks.

Two Notes of Caution
When Selecting Items Automatically

Automated item selection can have great advantages, but it should be applied with care and caution. The next two remarks serve to prepare researchers for applying the AISP to their data. First, before the AISP is applied, researchers are strongly advised to predict the most likely dimensional structure for their item set. Sometimes this can be done on the basis of the substantive theory underlying the construction of the items and sometimes on the basis of the content of the items. Explicit expectations may create an increased sensitivity to unexpected or odd selection outcomes and, in general, may also help to interpret the item clustering produced by the AISP. The idea is to think first, before the computer takes over. For example, a careful theoretical analysis based on the literature on transitive reasoning might reveal at the stage of item construction that the cognitive processes needed for solving inequality tasks are different from those needed for solving equality tasks. When this is known beforehand, the finding that the item set is not unidimensional can be put into better perspective.

Second, it may be noted that the AISP is not a formal test of the MHM. Because the AISP selects and rejects items on the basis of inter-item correlations (Equation 5.1) and item H_i coefficients (Equation 5.2), it sometimes may *select* an item that has an IRF showing a few local decreases, or *reject* an item with an increasing but relatively flat IRF. In these examples, the selected item is in *conflict* with the MHM assumption of monotonicity, and the rejected item is in *agreement* with the

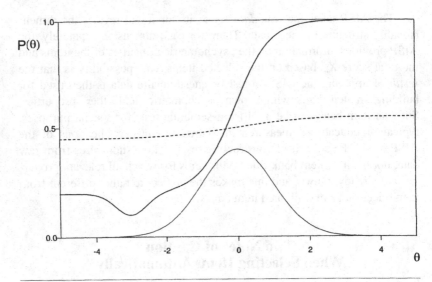

Figure 5.1   IRF (Solid Line) That Decreases Where the Frequency of the θs Is Low (Left) and Increases Where the Frequency of the θs Is High (Middle); and IRF (Dashed Line) That Is Nondecreasing and Almost Flat for All θs. *Note:* $f(\theta)$ is normal shaped.

monotonicity assumption. It is the choice of lowerbound c that may produce such results. This is explained next.

In the first case, items with locally decreasing IRFs may disturb the person ordering at the θ scale where the IRF decreases. In regions of the θ scale where the IRF increases, the item may contribute positively to person ordering. When the frequency of the θs is low where the IRF decreases and high where the IRF increases (see Figure 5.1, solid IRF), the H_i value may be high enough for item i to be included in the scale, especially when c is not high. From a *practical* point of view, for example, when the scale contains only few items and reliability of X_+ is low, it may be defended to include item i in the scale, even though it is not a good measure for *all* θs.

In the second case, items with almost flat but nondecreasing IRFs satisfy the MHM assumption of monotonicity (see Figure 5.1, dashed IRF). Although such items contribute positively to person ordering at all θ levels, they do so with considerable random error. From a *practical* point of view, the rejection of such model-fitting items, which may happen particularly when c is relatively high, may be defended because they add little to a reliable person ordering.

The previous discussion implies that the result of the AISP should not automatically be taken as final and that *manifest monotonicity* (Chapter 3) should be investigated for each item within each cluster separately. Fortunately, Equations 5.1 and 5.2 put so many constraints on the data that the selected items typically do not show many strong violations of manifest monotonicity. A serious *decrease* of an IRF in a region of the θ scale where *many* people are located will result in weak discrimination power, which usually is reflected by a low H_i value for that item. Such items are usually detected by the AISP. Thus, although the AISP is not a formal test for the MHM, it produces one or more item clusters that constitute an excellent *starting point* for further item analyses.

The main virtues of the AISP are that (a) it removes most of the items that do not satisfy the MHM, using Equations 5.1 and 5.2; (b) depending on the choice of lowerbound c, it uses Equation 5.2 to remove items that contribute hardly or only modestly to the reliable ordering of the respondents; and (c) it reveals the dimensionality structure of the data by selecting one or more subsets of items.

The Algorithm for Item Selection

This section discusses the AISP by distinguishing separate chronological steps. We use an item index j_s $(s = 1, 2, \ldots)$, where $s = 1$ indicates that the item is selected in the first step, $s = 2$ that the item is selected in the second step, and so forth. An example follows on page 73.

First Scale

Step 1. Mokken (1971, pp. 190–193) proposed an algorithm that, in the *fully automated version*, starts with the item pair (i, j_1), which has the largest H_{ij_1} of all item pair coefficients that are significantly higher than 0. In the *user-controlled version*, the researcher determines a start set of arbitrary size $(1 \leq k_{sel} < K)$. From that point on, both versions are identical. We will arbitrarily assume that the item pair with the highest significantly positive H_{ij} was selected, and move forward from that point.

Step 2. From the pool of the remaining $K - 2$ items, item j_2 is selected, which (a) correlates positively with both items i and j_1; (b) has an H_{j_2} with respect to the already selected items i and j_1 that is significantly higher than 0 and also higher than c; and (c) maximizes the total H for items i, j_1, and j_2. If several items meet the first two requirements, the third requirement facilitates a unique choice from among the items.

Step 3. In the third step, the item j_3 is selected from the pool of $K - 3$ remaining items, which (a) correlates positively with each of the items i, j_1, and j_2; (b) has an H_{j_3} with respect to the already selected items i, j_1 and j_2 that is significantly higher than 0, and also higher than c; and (c) maximizes the total H for items i, j_1, j_2, and j_3.

Next Steps. The procedure in Steps 2 and 3 is repeated until none of the remaining items satisfy (a) positive correlation with all items already admitted, and (b) an H_{j_s} with respect to these items that is significantly positive and also higher than c. Decision rules have been devised for special cases, for example, when two or more items produce the same maximum total H. Further, in each selection step, the Bonferroni inequality is used to adapt the significance level of the test of the null hypothesis, $H_{j_s} = 0$, to the number of significance tests in this and the previous steps. This means that in Step s, the desired overall significance level, say, 0.05, is divided by the number of tests performed in the previous $s - 1$ steps plus the number in Step s.

Next Scales

The AISP continues until no items remain that satisfy all three requirements. If any items are left, the AISP next attempts to select a second scale from these remaining items. This process is repeated until no additional scales can be formed. Notice that items that have already been selected in previous scales can no longer be considered for inclusion in later scales. The result is that an item occasionally is assigned to one scale but would fit better in another scale. To prevent this, the AISP can be rerun taking, for example, the second or third scale as a start set. This way, the researcher can consider several slightly different selection outcomes; this is also a measure to protect against uncritically accepting an outcome as true.

An Empirical Example: Analysis of Transitive Reasoning Data Revisited

The data set used in Chapter 3 to construct a preliminary test for transitive reasoning according to the MHM was reanalyzed using the AISP, which is contained in MSP. MSP uses a default lowerbound $c = 0.3$. We will also use $c = 0.3$ in this section and try alternative values of c with two other data sets in later sections.

Table 5.1 shows that the AISP produced two scales. The first scale is the final scale that was found in Chapter 3 after four rounds of careful data analysis using an informal top-down strategy. Note that a similar agreement of results will not always occur, because other strategies leading to different solutions could have been followed in Chapter 3. The second scale consists of the only two items (Items 2 and 4) that use equality relations only. We now look in more detail at the selection process for the first scale.

Step 1

Six selection steps were used to construct the first scale. Items 6 and 7 were selected first from the $K = 12$ items on the basis of $H_{6,7} = 0.78$. This was the highest H_{ij} of all item pairs that were significantly greater than 0 ($Z_{6,7} = 6.14$; $Z_{crit} = 3.20$, based on 66 item pairs [12 items] and an adjusted significance level of $0.05/66 = 0.0008$).

Step 2

In the second step, Equation 5.1 becomes active. On the basis of Equation 5.1, items that are negatively correlated with already selected items or, equivalently, items with negative H_{ij} values are rejected as candidates for selection into a scale. Here, this concerned the two pseudo tasks, Items 11 ($H_{6,11} = -0.27$) and 12 ($H_{7,12} = -0.28$), and Item 2 (equality relations only; $H_{2,7} = -0.03$), which were rejected at the start of the second selection round. This means that they played no further role in the construction of the first scale.

The item that was to become item j_2 was selected from the remaining seven candidates; it had the highest scalability value H with already selected Items 6 and 7, and also satisfied Equations 5.1 and 5.2. For each of the seven candidates, the total scalability coefficients of Items 6 and 7

Table 5.1 H_i Results of Automated Selection of Transitive Reasoning Tasks With Lowerbound $c = 0.3$.

SCALE 1				\multicolumn{6}{c}{Item H_i in Selection Step}					
Task	Property	Format	P_i	1	2	3	4	5	6
6	area	$Y_A > Y_B > Y_C$	.97	.78	.76	.66	.64	.64	.59
7	length	$Y_A > Y_B = Y_C$	.84	.78	.59	.56	.50	.51	.51
9	length	$Y_A = Y_B < Y_C = Y_D$	.30		.53	.60	.57	.51	.50
8	weight	$Y_A > Y_B = Y_C$	.97			.59	.62	.61	.55
3	weight	$Y_A > Y_B > Y_C$	.88				.51	.53	.53
10	weight	$Y_A = Y_B < Y_C = Y_D$	.52					.52	.52
1	length	$Y_A > Y_B > Y_C$	.94						.46
Total H				.78	.60	.59	.55	.53	.52

Rejected Tasks (Negative Covariance)   11   4
                                            12
                                             2

Excluded Tasks (Too Small/Nonsign. H_i)   5

SCALE 2				Item H_i in Selection Step
Task	Property	Format	P_i	1
2	length	$Y_A = Y_B = Y_C = Y_D$	.81	.35
4	weight	$Y_A = Y_B = Y_C = Y_D$	.78	.35

Rejected Tasks   None
Excluded Tasks (Too Small/Nonsign. H_i)   12
                                                      11
                                                       5

with each one of these seven items separately was calculated; that is, we have for the choices

j_2:	1	3	4	5	8	9	10
H:	0.53	0.52	0.14	0.21	0.59	**0.60**	0.59.

Thus, the total H of Item 9 with Items 6 and 7 was the highest. Also, Item 9 satisfied Equation 5.1: $H_{6,9} = 0.70$ and $H_{7,9} = 0.50$; and Equation 5.2 was satisfied: The item H_i value with the two already selected items was $H_9 = 0.53$ with $Z_9 = 3.26$. Thus, Item 9 was selected as the third item in the first scale. It may be noted that, in general, with each selection step the H_is change because they are based on the relation with the items selected in the previous step *and* the newly selected item. For example, in Step 1, $H_6 = H_{6,7} = 0.78$, and in Step 2 after the selection of Item 9, $H_6 = 0.76$.

Steps 3 Through 6

Item 4 (equality relations only) was rejected at this point as a possible candidate for future selection because $H_{4,9} = -0.12$ (Equation 5.1); this left only five candidate items for admission to the first scale. In the third selection step, a fourth item was selected from these remaining five items, and selection continued until *seven* items were selected. The only item left (Item 5: reversed format) was then *excluded* as a candidate for selection because it had a significantly positive but too low H_i ($H_5 = 0.20$ is lower than $c = 0.3$) with the other seven selected items (see Table 5.1). Table 5.1 shows that the total H value decreased during the selection process. This is a common finding, which is discussed in the next section.

Next, the AISP constructed a second scale containing Items 2 and 4 from the five items that were left over from the selection of the first scale, and *excluded* three of these items from this scale ($H_{12} = 0.12$, $H_{11} = 0.08$ and $H_5 = 0.05$). A third scale could not be produced, because the remaining Items 5, 11, and 12 had negative mutual H_{ij}s.

Selection Order of Items and Values of H During Selection

Theoretical research (Sijtsma, 1988) has shown that the AISP tends to select items measuring the same latent trait θ in a particular order according to their difficulty and that the overall H coefficient tends to decrease during item selection. It may be noted that the AISP does not *explicitly* select items according to their P values; as we have seen, it selects items according to Equations 5.1 and 5.2 and maximizes the H coefficient in each step. The results of this theoretical research are important for practical researchers, however, because they explain a pattern of scale results often found as a result of the AISP, and also make the results of item selection more predictable. The following rules of thumb hold by approximation when the items have difficulties that are widely spaced and the IRFs have slopes that are approximately equal. The distribution of the latent trait θ should be approximately normal. We will assume that the lowerbound $c = 0$, which is the minimum allowed under the MHM. This unusual choice allows for the selection of most items, making it easier to see what happens when the AISP can freely select items. Under these conditions, the following rules of thumb apply *in general*.

1. The first two items selected are the easiest and the most difficult in terms of the sample P_is. The third item selected has medium difficulty. The fourth and fifth items lie approximately halfway between the easiest and the medium items, and the medium and most difficult items. The next items are selected in such a way that "gaps" between already selected items on the latent difficulty scale are filled in evenly.

Because a sample H_{ij} based on an extremely low P_i and an extremely high P_j can be very unstable, even for large sample sizes, the null hypothesis that $H_{ij} = 0$ sometimes cannot be rejected and then these items are not selected first. When this happens, a less extreme item pair is selected first, but this does not happen very often. Of course, when the items in the complete item set do not have widely spaced difficulties, the zigzag pattern (to be illustrated shortly) in the P_is would not be as obvious. Further, when the original item set also contains some IRFs with relatively flat slopes, these items are selected last. Items with slopes that are relatively steep are often among the starting pair.

2. *H decreases* as the test length increases. This decrease tends to be fast with the first five items selected, and slow with the next items selected. For longer tests, H tends to level off to a positive value that depends on the discrimination of the items, given a fixed θ distribution; the higher the item discrimination, the higher the final H value.

These results are illustrated next by means of a real data example. Other interesting features of the AISP that are often encountered in practical item analysis are also discussed and illustrated using this data example.

An Empirical Example: Verbal Intelligence

Data Description

A data set of 0/1 scores obtained from 990 subjects who responded to 32 items measuring verbal intelligence can illustrate what happens to H in the course of the AISP, and how item difficulty is spread at each stage of the AISP. The subjects were at or beyond the college/university level and were tested in the context of personnel selection for computer-related jobs. The items were verbal analogies; for example:

_____ is to love as hostility is to _____

1. kiss   2. enemy   3. wedding   4. hate   5. lover

A. ally   B. law   C. quarrel   D. passion   E. friendship

The first missing word has to be chosen from the first row, and the second missing word from the second row, such that a correct analogy appears. Item numbers 1 through 32 reflect the presentation order, not the difficulty ordering.

Start Set for Item Selection

Figure 5.2 reveals that the start set did not consist of the easiest item, $3(P_3 = 0.93)$ and the most difficult item, 31 $(P_{31} = 0.14)$. For these items, the item pair scalability coefficient (Chapter 4) equals $H_{3,31} = 1 - F_{3,31}/E_{3,31} = 1 - 4/9.5 = 0.58$ $(Z_{3,31} = 1.98)$, which was not significant at a level adapted to the large number $(31 \times 32/2 = 496)$ of H_{ij}s. Due to low expected error frequency, H_{ij} can be very unstable if one item has a small P_i and the other a large P_j. This instability is reflected in a low Z_{ij} value. Z_{ij} is a conservative test in this situation, which can be considered a desirable property. Instead, the item pair (21, 28) was the start set. This item pair had the highest significant H_{ij} value: $H_{21,28} = 0.65$, with $Z_{21,28} = 8.99$.

Item Selection

With $c = 0$ as lowerbound, 31 items were stepwise selected into one scale. Item 22 was rejected due to negative scalability coefficients with Items 3, 6, and 12. For the 31 selected items, the distribution of X_+ had a mean of 16.94, a standard deviation of 6.32, a skewness of −0.11, and

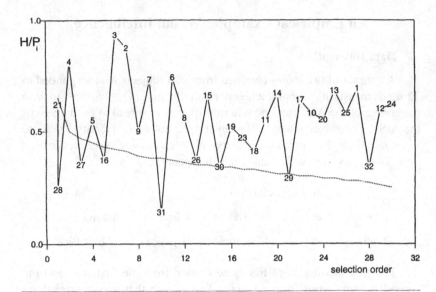

Figure 5.2   Selection Order of 31 Verbal Analogy Items and Development of overall H (Dotted Curve). Item numbers represent P_i sample values. *Note:* Item numbers represent original item numbering from test.

a kurtosis of –0.81. Thus, the distribution was almost symmetrical. The P_i values of the 31 items ranged from 0.14 to 0.93, and 18 items had P_is between 0.50 and 0.75. Thus, the P_is were widely spaced, although not equidistant; this is realistic for many empirical data sets. The H_is ranged from 0.12 to 0.39, and 23 of these H_is were between 0.2 and 0.3. This indicates that the slopes of most IRFs probably showed little variation. An almost symmetrical X_+ distribution, widely spaced P_is, and slopes showing little variation are in agreement with the conditions under which the rules of thumb for the selection results of the AISP hold.

Figure 5.2 shows the zigzag pattern of item popularities P_i when selecting items by means of the AISP. The dispersion of the P_is was largest for the first 11 or 12 items. The remaining items had P_is with much smaller dispersion, and the selection pattern thus became less jumpy. Of course, the exact pattern depends on the spread of the P_is of this item set.

Figure 5.2 shows that H tended to decrease as predicted: After a fast decrease with the selection of the first few items, H decreased only slowly during the selection of the next items. In Figure 5.2, it may be noted that H did not really approach an asymptote value. The reason is that because the AISP always selects the "best" item from the remaining $K - k_{sel}$ "worst" items, and realistic item pools consist of items with at least some variation in quality (discrimination), it may be expected that in practice, H will continue to decrease. This decrease is expected to be slower as the variation in item quality becomes smaller.

Choice of Lowerbound c

The choice of lowerbound c has a strong influence on the composition of the selected scale or scales. For $c = 0$, the only effective constraint on the end result is that all selected items within the same scale have positive inter-item correlations (or, equivalently, $H_{ij} > 0$). This led to the selection of 31 out of 32 verbal analogy items. For higher c values, items must have IRFs with steeper slopes to be selected into the same scale. The result of a higher c tends to be that (a) fewer items are selected into a given scale, but these items tend to have high discrimination; (b) a larger number of shorter scales is formed; and (c) more items turn out to be unscalable; that is, the *total number* of items selected into different scales is smaller. For our example, for $c = 0.3$, 17 items were selected into the *first* scale ($H = 0.34$) and 3 items into the *second* scale ($H = 0.32$) (two scales selected in total); for $c = 0.4$, the *first* scale contained 6 items ($H = 0.43$) and the *second* through *fifth* scales each contained 2 items (*H*s equal to 0.58, 0.56, 0.47 and 0.43; five scales selected in total); and for $c = 0.5$, five 2-item scales were selected (*H*s equal to 0.65, 0.58, 0.56, 0.55, 0.51).

Because all 32 items in the original item pool had the same formal structure, and because there was no evidence of multidimensionality, our choice of the final solution will be between the solutions for $c = 0.0$ and $c = 0.3$, and may depend on technical considerations such as reliability of the total score X_+ and available testing time. The solutions for $c = 0.4$ and $c = 0.5$ were the results of a scaling criterion that was too high and that caused the item pool to crumble into meaningless small subscales, with too many items excluded.

Too Low Final Item Scalability

Sometimes a scale resulting from the AISP contains one or two items with scalability H_i values that are lower than the lowerbound c. Such H_i values are in conflict with Equation 5.2 of the definition of a scale, which requires items to have $H_i \geq c$. For example, for $c = 0.3$, it was found that $H_{31} = 0.28$; this is the H_i of Item 31 with respect to the other 16 items selected. It is crucial to note, however, that Item 31 was selected as the 11th item and *at that moment* had an item scalability of $H_{31} = 0.31$ with respect to the 10 already selected items (these values can be retrieved from the MSP output). This value was *higher* than $c = 0.3$; thus, Equation 5.2 was satisfied. With the selection of the *next* (12th) item, H_{31} dropped to 0.30, which was still sufficiently high, and H_{31} kept this value through the selection of the *next* 3 items. Then Item 30 was the 16th item to be selected, with item pair $H_{30,31} = 0.16$, which caused H_{31} to drop to 0.29; and Item 19 was the 17th item to be selected, with item pair $H_{19,31} = 0.17$ and H_{31} finally dropped to 0.28.

Note that although in the final scale Item 31 did not satisfy the definition of a scale, it did when it was selected. One might decide, of course, to overrule the AISP and delete Item 31 from the scale. However, another decision might be that because all items are so much alike in content, Item 31 could be included to increase the reliability of X_+.

Selection of Items From a Multidimensional Item Pool

Test constructors often want their test or questionnaire to measure several aspects of the latent trait under consideration. For example, when constructing a questionnaire for the measurement of coping when confronted with a particular problem, a researcher may decide that coping has several aspects and that the most important ones should be represented in the questionnaire. The usual strategy is to construct several items for each of the aspects of coping to be included in the questionnaire. The result of this strategy may be that the data collected by means of the questionnaire are *multidimensional*. This multidimensionality may be revealed by means of the AISP. If different subscales are found, the researcher may decide to give persons a separate total score on each subscale rather than one total score on the entire item set.

The AISP is able to find the true dimensionality structure of the data when the appropriate lowerbound c is chosen. If subsets of different items measure different latent traits, this *multidimensionality* may be obscured by the AISP if lowerbound c is either too low or too high. If items measuring different traits have moderate inter-item correlations, c values close to 0 may allow all or almost all items into a single scale, whereas high c values may lead to the rejection of all items or cause the item pool to crumble into a few small scales because most items are excluded. These examples show that the choice of lowerbound c is important for establishing the true dimensionality.

Hemker, Sijtsma, and Molenaar (1995) ran the MSP software on several simulated data sets to find out to what degree the AISP was capable of finding the correct dimensionality according to which the data were simulated. *Ideally*, the AISP should select items measuring the same trait into the same cluster, and items measuring different traits into different clusters. They used as a *criterion for success* that the AISP selected at least two thirds of the items correctly. Repeating this for different c values, Hemker et al. (1995) concluded that there exists no unique c value or unique interval of c values that yield the correct or almost correct solution for all data sets. They gave the following recommendation. For empirical data sets, the AISP should be run repeatedly, starting with $c = 0$ and increasing c in each subsequent run by 0.05, until $c = 0.55$ is reached. Higher c values did not provide information about the dimensionality of the item pool. For unidimensional item pools and multidimensional item pools, different typical patterns of outcomes were found across the 12 solutions between $c = 0.0$ and $c = 0.55$.

In unidimensional item sets, the typical sequence of outcomes for increasing c is

1. most or all items are in one scale;

2. one smaller scale is found; and

3. one or a few small scales are found and several items are excluded.

The solution found at the first stage should be taken as the unidimensional scale. This means that for the verbal analogy items from the previous subsection, the 31-item scale found at $c = 0.0$ is the unidimensional scale. This choice of c may seem to be at odds with earlier recommendations about the lowerbound of $c = 0.3$, but the point here is that we use a whole range of c values, from 0.0 to 0.55, and use the pattern of cluster outcomes to make the final decision about the dimensionality of the data.

In multidimensional item sets, the typical sequence of outcomes for increasing c is

1. most or all items are in one scale;

2. two or more scales are formed; and

3. two or more smaller scales are formed and several items are excluded.

The scales found at the second stage should be taken as the final multidimensional solution. It may be noted that the difference between

Table 5.2   Short-hand Text of 17 Items Measuring Coping With Industrial Malodors, and Their Difficulty P_i.

Item Number	Item Text	P_i
1	Keep windows closed	.60
2	No laundry outside	.43
3	Search source of malodor	.59
4	No blankets outside	.43
5	Try to find solutions	.20
6	Go elsewhere for fresh air	.09
7	Call environmental agency	.05
8	Think of something else	.16
9	File complaint with producer	.06
10	Acquiesce in odor annoyance	.52
11	Do something to get rid of it	.22
12	Say, "It might have been worse"	.29
13	Experience unrest	.14
14	Talk to friends and family	.22
15	Seek diversion	.16
16	Avoid breathing through the nose	.17
17	Try to adapt to situation	.69

outcomes for unidimensional and multidimensional data sets lies in the second and third stages of the sequence of outcomes described above. An illustration of how a decision is made in practice follows next, when we apply the AISP to select unidimensional scales from a multidimensional item pool.

An Empirical Example:
Coping With Industrial Malodors

Data Description

Cavalini (1992) used a questionnaire to investigate how people living in the vicinity of a malodorous factory cope with industrial malodors. The questionnaire contained 17 items (Table 5.2), each with four ordered answer categories. Each item is a possible reaction to the question: What do you do or think when you smell malodor? The answer categories ranged from *never* (score 0) to *almost always* (score 3).

In a sample of 828 respondents, several factor analysis solutions were found (Cavalini, 1992, pp. 53–54); the four-factor solution gave the best interpretation. This led to four scales: Scale 1 (7 items: 3, 6, 8, 13–16) measured a mixture of emotional and avoidance reactions; Scale 2 (4 items: 5, 7, 9, 11) measured a rational effort to do something about the malodor problem; Scale 3 (3 items: 1, 2, 4) measured the effort to protect the inside of the house and the laundry from the bad outside air; and Scale 4 (3 items: 10, 12, 17) measured emotional acceptance of the inconvenience.

Dimensionality Analysis

Analysis of the polytomous item scores is postponed until Chapter 8. Here, we reanalyzed the *dichotomized* version of the polytomous data with 0 and 1 replaced by 0, and 2 and 3 by 1 (sample P_is in Table 5.2; item numbering does not correspond to increasing P_i). Table 5.3 shows the pattern typical of multidimensionality: (a) For c ranging from 0.00 to 0.20, most items were selected into one scale; in this case, the majority of the items were selected into Scale 1, and a minority into Scale 2; (b) for $c = 0.25$, three scales were found; and for $c = 0.30$, four scales were found; (c) for $c \geq 0.35$, only small scales were found; as c increased, the number of small scales decreased and more items were rejected or excluded. Because the rule of thumb says that we should accept the results of the second stage as final, we now further discuss these results and decide whether or not there is any reason to propose amendments.

We compare the results (dichotomized data) found using the AISP with Cavalini's factor analysis results (rating scale data), not because we expect to find exactly the same results, but mainly to better facilitate the interpretation of our results and to point out differences with factor analysis solutions. The solution with three scales ($c = 0.25$) had as its *first* scale a composite of the complete Scale 2 found by Cavalini (1992), four items (Items 3, 6, 13, 14) from his Scale 1, and one item (Item 1) from his Scale 3. It might be argued that Items 1, 3, and 6 represent a rational response and may be included with Items 5, 7, 9, and 11 to form one scale. Items 13 and 14 seem less suitable for this scale. Our *second* scale consisted of only Items 2 and 4; together with Item 1, they form Cavalini's Scale 3. Our *third* scale consisted of Cavalini's Scale 4 plus Items 8 and 15 from his Scale 1. Based on item contents, it could be argued that this combination makes sense without altering the interpretation of Cavalini's Scale 4.

Table 5.3 Stepwise Determination of Coping Behavior Scales Using the AISP.

			Item Numbers		
c	Scale 1	Scale 2	Scale 3	Scale 4	Scale 5
0.00	1–9, 11, 13, 14	10, 12, 15, 17			
0.05	1–9, 11, 13, 14	10, 12, 15, 17			
0.10	1–7, 9, 11, 13, 14	8, 10, 12, 15, 17			
0.15	1–7, 9, 11, 13, 14	8, 10, 12, 15, 17			
0.20	1–7, 9, 11, 13, 14	8, 10, 12, 15, 17			
0.25	1, 3, 5–7, 9, 11, 13, 14	2, 4	8, 10, 12, 15, 17		
0.30	3, 5–7, 9, 11	1, 2, 4, 13	8, 15, 17	10, 12	
0.35	3, 5, 7, 9, 11	1, 2, 4	8, 15, 17	13, 14	10, 12
0.40	3, 5, 7, 9, 11	1, 2, 4	8, 17	13, 14	10, 12
0.45	3, 5, 7, 9, 11	1, 2, 4	8, 17	13, 14	10, 12
0.50	3, 5, 7, 9, 11	1, 2, 4	8, 17	13, 14	10, 12
0.55	7, 9, 11	2, 4	3, 5	1, 6	
0.60	9, 11	3, 5, 7	2, 4		

Table 5.4   Detailed Item Selection Results of Coping Data for Lowerbound $c = 0.30$ and $c = 0.35$.

		Lowerbound				
		0.30			0.35	
Scale	Item	H_i	H	Item	H_i	H
1	3	.57		3	.60	
	5	.51		5	.53	
	6	.30				
	7	.53		7	.60	
	9	.50		9	.63	
	11	.45	.47	11	.49	.55
2	1	.52		1	.54	
	2	.52		2	.68	
	4	.61		4	.67	
	13	.30	.55			.64
3	8	.41		8	.41	
	15	.38		15	.38	
	17	.49	.42	17	.49	.42
4	10	.47		13	.53	
	12	.47	.47	14	.53	.53
5				10	.47	
				12	.47	.47

The solution with four scales ($c = 0.30$) seems to be more fortunate than the three-scale solution. Our *first* scale was a subset from the corresponding first scale for $c = 0.25$: The misfitting Items 13 and 14 and Item 1 were excluded. For the remaining six items, Table 5.4 shows that Item 6 had $H_6 = 0.30$, which is much lower than the scalability values for the other five items. Also, Item 6 expresses a coping strategy ("go elsewhere for fresh air") that is less active than each of the strategies expressed by the other five items. Using $c = 0.35$ led to the removal of Item 6, leaving a five-item subset that had a more homogenous content and higher scalability values. This scale almost coincided with Cavalini's second scale. For $c = 0.30$, Item 1 was part of our *second* scale, together with Items 2, 4, and 13. Although Item 13 did not really disturb the meaning of this scale, its $H_{13} = 0.30$, which was much smaller than the scalability values of the other three items. Again, using $c = 0.35$ led to the removal of this item, leaving a more clear-cut interpretation for the remaining scale and higher scalability values for each item. This second scale was identical to Cavalini's third scale. Our *third* scale ($c = 0.30$) was a part of Cavalini's first scale. Finally, our *fourth* scale ($c = 0.30$) contained two of the three items from Cavalini's fourth scale.

Based on meaning alone, our four-scale solution ($c = 0.30$) is better interpretable than the three-scale solution ($c = 0.25$). Moreover, it preserves the meaning contained in Cavalini's four factors. However, the composition of the first two scales is even better for $c = 0.35$ than for $c = 0.30$, and the scalability values for individual items and for the whole scale are higher. Thus, we might decide to accept the first two scales for $c = 0.35$ as final, together with the third and fourth scales for $c = 0.30$. This conclusion shows that an item analysis may involve a rather complex decision process that combines statistical and substantive considerations in an effort to find a satisfactory solution.

Discussion

The automated bottom-up item selection procedure can be of great help to researchers, because (a) it replaces a time-consuming top-down procedure that in practice is done "by hand"; (b) it prevents the dimensionality structure of the item set from being overlooked in the process of rejecting items step by step or in small clusters; and (c) it protects against chance capitalization due to decisions based on the high number of significance tests at the different stages of the selection process.

One might argue that the dimensionality structure of the item set also can be investigated using factor analysis. Factor analysis of discrete item scores, in particular dichotomous item scores, is hazardous because product-moment correlations are lowered to a greater degree the more the P_is of the two items differ (see Chapter 4 for an illustration). Thus, one cannot be sure whether a low correlation between items with different P_is is due to this ceiling effect, or to different traits underlying the responses, or both. The use of tetrachoric correlations alleviates problems due to the ceiling effect, but introduces new problems because such correlations tend to overestimate the strength of the relationship between the items. The AISP, which is based on a sequential cluster analysis, circumvents these problems because it uses the H coefficient as a criterion for including items in a scale: Because H is a weighted sum of covariances normed against the weighted sum of *maximum possible covariances given the P_is*, the annoying ceiling effect is absent.

The full application of the AISP in exploratory item analysis requires several runs using different lowerbound cs for H to find the dimensionality structure of the items. This calls for judgment on the part of the researcher. We emphasize once more that the final outcome of automated item selection has to be approved of by the researcher, which can ultimately mean that the outcome is rejected. Also, note that the definition of a scale (Equations 5.1 and 5.2) is mathematical: It does not guarantee an operational relation of the measured variable to the intended hypothetical construct; in other words, it does not guarantee construct validity. The AISP is based on this mathematical definition and thus mechanically selects items into clusters in agreement with this definition. Hence, the AISP cannot be a substitute for a theoretically meaningful analysis of the value of the instrument for social science research. The AISP does not eliminate the need for a final intuitive, theoretical, and commonsense evaluation of the scale and its items on the basis of their content. The final value of the scale stems from its construct validity: from the theoretical significance of the way in which the scale relates the variable it is supposed to measure to other variables in a broader operational context (Mokken, 1971, p. 194).

Additional Reading

Many researchers (e.g., Cronbach, 1951; Horst, 1953; Guilford, 1954; Cliff, 1977; Terwilliger & Lele, 1979; Cudeck, 1980; Raju, 1982) have studied differences and similarities between the H coefficient and the well-known lowerbound to the reliability, alpha (e.g., Cronbach, 1951). H also has been investigated as a function of the tetrachoric correlations between the items (Cronbach, 1951; Carroll, 1961) and the item P_is (Jansen, 1982a, 1983). Both H and alpha have been studied as a function of the mean and the variance of the item difficulties, the discriminations, and the pseudo-guessing parameters of logistic IRFs, and the mean and the variance of θ (Cudeck, 1980); varying distributions of the test score and the difficulty level of the test (Terwilliger & Lele, 1979); and the sum of proportions of Guttman errors across item pairs and the item difficulty of 1PLM IRFs (Sijtsma, 1988, pp. 89–94). From a review of this research (Sijtsma, 1988, pp. 85–95), it was concluded that H and alpha should not be used interchangeably.

The AISP has been described by Mokken (1971, chap. 5; also see Mokken & Lewis, 1982; Sijtsma, 1998). The selection order of items from an item pool by means of the AISP was studied by Sijtsma and Prins (1986). Hemker et al. (1995) studied the use of the AISP for investigating the dimensionality of an item pool. Van Abswoude, Van der Ark, and Sijtsma (2001) compared the AISP with the item selection algorithms DETECT (Zhang & Stout, 1999) and HCA/CCPROX (Roussos, Stout, & Marden, 1998). Another method was proposed by Bolt (2001). The verbal intelligence data were previously described and analyzed by Meijer, Sijtsma, and Smid (1990).

Exercises for Chapter 5

5.1. Summarize in your own words the main advantages of an automated item selection procedure.

5.2. For stepwise increasing values of the lowerbound c, one expects one of two sequences of outcomes listed as 1, 2, and 3 at the end of the subsection called "Selection of Items From a Multidimensional Item Pool." Which sequence is observed depends on whether the item pool was unidimensional or not. Provide a motivation why these two sequences will probably occur.

5.3. Table 5.5 contains the H_{ij}s, P_is, and the H_is for six items (item contents in Table 5.2).
   a. Which item pair do you think is selected first by the AISP? Do you have any reservation?
   b. What comment can you make on the P_is of the item pair you chose in Exercise 3a?

Table 5.5

Item Number	H_{ij}						P_i	H_i
	13	8	15	4	1	17		
13	—						.14	.16
8	02	—					.16	.24
15	.25	.36	—				.16	.30
4	.27	.10	.11	—			.43	.26
1	.40	.38	.43	.54	—		.60	.32
17	−.18	.55	.43	.10	.05	—	.69	.14

    c. Given the H_{ij}s and the H_is in the table, would you actually expect these items to form one scale?

    d. If all six items were considered as one scale, how high would H be, at the least?

5.4. Table 5.6 shows the results of running the AISP with lowerbound $c = 0.3$ on the six items from the previous exercise.

    a. Given the information in this table and Table 5.5, which item seems to fit least convincingly into one of these two scales?

    b. Give the boundaries for the overall H of the two scales.

    c. Can you be sure that each of the two scales has a higher overall H than the scale based on all six items?

Answers to Exercises for Chapter 5

5.1. The AISP removes most of the items that do not satisfy the MHM, and for suitable choices of the lowerbound c, also items that contribute little to reliable respondent ordering. Moreover, it discloses which items cluster together in cases where multidimensionality is present in the item pool.

5.2. In unidimensional item pools, pairwise population H values will all be positive, so for low enough c values, one expects all items in the scale (sampling fluctuation or lack of significance may cause exceptions). If c rises above the lowest item H value thus obtained, that item is no longer a candidate for selection. When c rises further, more items drop out. For still higher c values, only occasional small clusters have items with a mutual association that is strong enough that the selection criteria are met, and the other items remain isolated.

For multidimensional item pools, the first and the last stage are similar, but at the middle stage, a single scale consisting of most items is not expected but rather two or more meaningful subscales, because now H

Table 5.6

Scale 1		Scale 2	
Item Number	H_i	Item Number	H_i
8	.41	1	.33
15	.38	4	.46
17	.49	13	.51

values within these subscales will generally be higher than H values between items belonging to different subscales.

5.3. a. Without statistical testing, item pair (8, 17) because it has the highest H_{ij}.

    b. Item 17 is the easiest, and Item 8 almost the most difficult item. Given the somewhat smaller P value of Item 13, one would have expected this item to be selected first together with Item 17; however, $H_{13,17} = -0.18$.

    c. The H_is are quite low. Moreover, the H_{ij}s show a great amount of dispersion. This may suggest multidimensionality.

    d. $H \geq 0.14$.

5.4. a. Item 1. The H_i value of this item deviates more than any other H_i value from the other two H_i values in the same scale. Moreover, based on H_{ij} values, Item 1 seems to scale rather well with Items 8 and 15 from the first scale. However, it scales badly with Item 17.

    b. Scale 1: $0.38 \leq H \leq 0.49$; Scale 2: $0.33 \leq H \leq 0.51$.

    c. Yes; for both three-item scales, the lower bounds for H are higher than the upperbound for the total item set, which equals 0.32. The actual H values reported by MSP were 0.42 (Scale 1), 0.45 (Scale 2), and 0.24 (all six items).

6

The Double Monotonicity Model

Invariant Item Ordering

The double monotonicity model (DMM) was introduced in Chapter 2. Like the monotone homogeneity model (MHM), the DMM is based on the assumptions of unidimensionality, local independence, and monotonicity. Unlike the MHM, however, the DMM in addition assumes that the IRFs do not intersect. Nonintersection is an assumption that also has been called invariant item ordering (IIO). A set of k dichotomous items is said to exhibit an IIO if the items can be ordered—and numbered accordingly—such that

$$E(X_1|\theta) \leq E(X_2|\theta) \leq \ldots \leq E(X_k|\theta), \text{ for each } \theta, \quad (2.14)$$

or equivalently,

$$P_1(\theta) \leq P_2(\theta) \leq \ldots \leq P_k(\theta), \text{ for each } \theta \quad (2.8)$$

Ties may occur for some values or intervals of θ.

Because Equations 2.14 and 2.8 condition on θ, it is clear that an IIO applies to individual θs. This means that an IIO can be an important property in applications where tests are used for comparing persons with respect to their performance on individual items. An IIO also implies that the item ordering is invariant across different subgroups from the population of interest. This was explained in detail in Chapter 2. In particular, it was shown that the IIO in Equation 2.14 implies, by averaging the conditional expectations over the distribution of θ (which means integrating, because θ is continuous), that in the *whole population* of interest, the proportions-correct, the P_is, have the same ordering as the conditional expectations:

$$P_1 \leq P_2 \leq \ldots \leq P_k. \tag{2.19}$$

Chapter 2 also showed that this result held regardless of the exact form of the distribution of θ. In other words, in the DMM, the item *ordering* by means of the P_is is the same, with the exception of possible ties, in each subgroup as well.

The reader may have noted that we are talking explicitly about the *ordering* of the items, not about the *exact values* of the Ps. Of course, the exact value of a particular P_i, say, P_1, may vary across subgroups because it depends on the distribution of θ, and this distribution may vary across subgroups. That is, P_1 may be 0.15 for boys and 0.27 for girls; but under an IIO, in both subgroups, Item 1 is the most difficult item relative to the other items. Thus, $P_2 \geq 0.15$ for boys and $P_2 \geq 0.27$ for girls, and so on. The important point is that an IIO (Equation 2.14) implies the *same ordering* in all subgroups. This result proves to be useful in several applications, as we see next.

Practical Application of Invariant Item Ordering

An IIO is an important property when individuals or groups are compared. We give practical examples below of when the invariant ordering of the items is important for individual diagnosis or the comparison of groups. In which situations would an IIO be desirable?

Testing Psychological Theories. IRT models can be used to test theories about psychological constructs. For example, it may be hypothesized that an ability develops along a fixed sequence of stages. The developmental theory about this ability might predict the characteristics of a task that a child at that stage should be able to solve. Following this idea, it may be possible to construct a set of tasks, each of which corresponds to another stage of the developmental process and requires different subabilities for its solution. A test containing these tasks would be highly useful for the diagnosis of a child's developmental stage.

Because the theory claims a general, fixed sequence of developmental stages, and the items tap subabilities corresponding to these stages, we expect an ordering of difficulty of the items that corresponds to the developmental stages, and that is the same for all children. We use the developmental ability of transitive reasoning to make up an example based on the items discussed in Chapter 3. We might hypothesize that tasks about equality relationships are more difficult the more objects that are involved. Also, tasks about inequality relationships may be hypothesized to be more difficult than tasks involving equality relationships; and inequality tasks involving weight problems may be hypothesized to be more difficult than inequality tasks involving length problems; and so on.

A test containing such tasks could be administered to children at different developmental stages. A child solving only equality tasks would be diagnosed to be at an earlier stage than a child who also solves inequality tasks about length and weight. Clearly, in order to draw conclusions about a child's developmental level on the basis of such tasks, an IIO should hold for the test in the population of interest.

Starting and Stopping Rules. When testing cognitive abilities and educational achievements, the items often are administered in an ascending difficulty ordering. One reason for this presentation order is to prevent the tested child or student from panicking if the first items happen to be difficult and, as a result, achieving far under his or her normal level on the remainder of the test. Another reason is to administer to an examinee only those items that are neither too easy nor too difficult, and this is accomplished by using starting and stopping rules in combination with an ascending difficulty order. The next example clarifies the use of such rules.

In intelligence testing, for a population of 6- to 12-year-olds, the same item ordering may be used together with different starting and stopping rules for each age group. The youngest age group starts with the first item, and each child stops when it fails, say, at three consecutive items. Because the items are ordered from easy to difficult, we may assume that the next items will also be failed with high probability because they are more difficult than the items already attempted. The next age group skips, say, the first five items because these items are too easy for them and starts at the sixth item. The same stopping rule applies. The third age group starts at, say, the 13th item, and individuals again stop using the same stopping rule, and so forth. Examples of intelligence tests that use this administration procedure are the Revised Amsterdam Child Intelligence Test (Bleichrodt, Drenth, Zaal, & Resing, 1985), the Snijders-Oomen nonverbal intelligence test (Laros & Tellegen, 1991), and the Wechsler Intelligence Scale for Children (Wechsler, 1999). Because intelligence tests are usually used for *individual* diagnosis, and because it is assumed that the item ordering applies to each individual, this administration procedure makes more sense if it is based on an IIO.

Aberrant Item Score Patterns. Researchers sometimes are interested in the pattern of 0s and 1s in addition to the total score X_+ or another estimate of θ. Methods to investigate whether item score patterns are atypical under a particular IRT model are known as appropriateness measurement methods or person-fit methods (Meijer, 1996). For example, patterns containing many 0 scores on easy items and at the same time many 1 scores on difficult items are suspect, because most IRT models predict that the probability of a 1 score is lower when items are more difficult. Thus, it may be possible that respondents who produce such item score patterns cheated on the most difficult items or were not concentrating when trying the easier items. Identification of such an examinee may be useful because his or her X_+ may overestimate θ (cheating) or underestimate θ (lack of concentration). This is useful information when using test scores for individual diagnosis and decision making.

For individual decisions based on item score patterns, it is convenient that the ordering of the items is the same for all individuals taking the test. That is, the interpretation of the test performances of two individuals is much easier when one of the test's properties is IIO. What happens when IIO does *not* hold? Let us consider a five-item test whose IRFs are shown in Figure 6.1. Suppose that for John with latent trait value θ_{John}, the item ordering in Equation 2.8 holds, and thus for John, $P_5(\theta_{John})$ is the largest probability of all five response probabilities. Also suppose that for Mary with $\theta_{Mary} > \theta_{John}$, due to the intersection of the IRF of Item 5 with several other IRFs between θ_{Mary} and θ_{John}, Item 5 is *not* the easiest item; that is, she has response probabilities that are larger than $P_5(\theta_{Mary})$ (Figure 6.1). Suppose that, like John, Mary also has Item 5 incorrect, but note that for her this item is more difficult relative to the other items. Thus, we have the situation that in the same test, John's having Item 5 incorrect may be suspect, but Mary's having Item 5 incorrect may not be an indication of aberrant response behavior. This situation occurs because the IRFs intersect. The DMM has nonintersecting IRFs and thus implies an IIO. Obviously, interpretation of person-fit results is easier if an IIO holds within the population of interest and, thus, for each of its members.

It may be noted that several person-fit methods, especially when they are based on the 2PLM or the 3PLM, do *not* assume an IIO. In cases where person-fit indices indicate that a particular item score pattern is

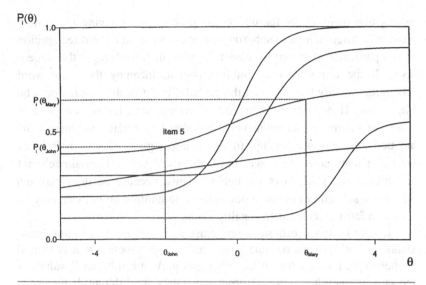

Figure 6.1   Five IRFs Under the MHM. Item 5 is the easiest for John and third-easiest for Mary.

aberrant, this may be taken as a warning that for that examinee the θ value is either under- or overestimated. This is useful information, of course, but when it comes to the interpretation of test performance in terms of item score patterns, the fact that an item may be relatively easy for one examinee but difficult relative to other items for another examinee makes interpretation awkward. Under the 2PLM and the 3PLM, interpretations of item score patterns thus have to be related to the item ordering at a particular interval of θ. With many intersections, it can be awkward to understand from a substantive point of view why item i is more difficult than item j for someone with a θ value in one interval, and easier for someone with a θ value in the next interval. Thus, an IIO facilitates interpretation of test performance.

Differential Item Functioning. Differential item functioning (DIF) occurs when the IRF of a particular item is different in two relevant subgroups taken from the population of interest. Usually, these subgroups are referred to as the reference group and the focal group $\frac{1}{M}$; for example, a white and a black group, boys and girls, or full-time workers and part-time workers. When the IRF of an item is different in the two subgroups of interest, this means that there are θs for which the probability of giving the correct answer is different in both subgroups. For example, in a word recognition test, a particular word may be easier for girls than for boys at the same θ level. If the test were truly unidimensional, meaning that only word recognition ability θ determined the probability $P_i(\theta)$, this result would be impossible. Hence, for the situation described here, the responses to at least a few items are determined also by other latent traits, and these traits are tied to group membership. In the example given, due to their faster cognitive maturation, girls may have already developed certain relevant subabilities on which boys lag behind. Thus, because an item may tap additional subabilities in one group relative to another group, DIF may be seen as a form of multidimensionality.

It may be noted that situations exist in which the test is *unidimensional* in both groups, but one group has a θ distribution that is shifted farther to the left than that of the other group. As a result, the P_i values of the same item evaluated in each group separately are different: In the group with the θ distribution farther to the left, the P_i value is lower than in the other group. Also, two persons with the same θ but from different groups may have the same response probability, $P_i(\theta)$, but one person have a relatively high position in his or her "parent" θ distribution (the one farthest to the left), and the other person have a lower position in the other "parent" θ distribution. The point we want to emphasize here is that the shift of the θ distributions between groups may appear to produce DIF because items have different P_i values between groups, but in fact, measurement is unidimensional and consequently all we are observing is a shift in θ level.

So far, we have discussed DIF (or lack of it) assuming that we know the latent mechanism, be it multidimensionality for some of the items (true DIF) or a simple shift of θ distributions (apparent DIF). In practice, we do not know these latent mechanisms, and DIF, if present, may cause the ordering of items with respect to their P_i values to be different in subgroups that are demographically, socially, or psychologically different. As noted earlier (Chapter 2), different item orderings by P_i values in such groups often raise the question of what caused these differences. This is particularly true if different groups together constitute the population in which the test is used to make decisions that simultaneously involve individuals from these different groups. Examples are educational admission and job selection on the basis of test scores. A simple inequality of item orderings in different groups may serve to put the researcher on the rail of DIF, and an IIO investigation may reveal this inequality.

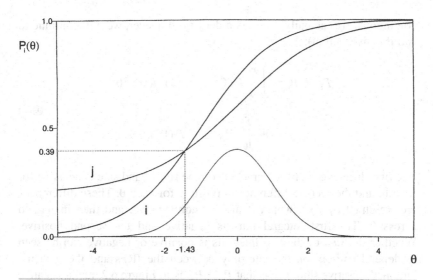

Figure 6.2   Two IRFs Intersecting at $\theta^* = -1.43$; and a Normal-Shaped Distribution of θ.

Investigating IIO

Figure 6.2 shows two *intersecting* IRFs and a distribution of θ, in the sequel denoted $f(\theta)$. The IRFs intersect at $\theta^* = -1.43$. To the left of θ^*, item j is easier than item i, and to the right, the order is reversed:

$$\text{For } \theta < \theta^*, P_i(\theta) < P_j(\theta); \tag{6.1a}$$

and

$$\text{for } \theta > \theta^*, P_i(\theta) > P_j(\theta). \tag{6.1b}$$

If the IRFs have more points of intersection, the ordering of the IRFs reverses each time an intersection point is passed.

For the whole group, the ordering of the overall proportions P_i and P_j depends on the interplay between (a) the IRFs $P_i(\theta)$ and $P_j(\theta)$, and (b) the distribution $f(\theta)$. For the situation in Figure 6.2, the next formula further clarifies the dependence of the ordering of P_i and P_j on the IRFs, $P_i(\theta)$ and $P_j(\theta)$, and the θ distribution, $f(\theta)$ (because θ is continuous, $f(\theta)$ actually is called a density; however, we will continue to call it a distribution):

$$
\begin{aligned}
P_i - P_j = &\int_{-\infty}^{\theta^*} [P_i(\theta) - P_j(\theta)] f(\theta) \, d\theta \\
&+ \int_{\theta^*}^{\infty} [P_i(\theta) - P_j(\theta)] f(\theta) \, d\theta.
\end{aligned}
$$

(6.2)

The first difference between brackets on the right-hand side is negative for each θ, and the second difference is positive for each θ. These differences are weighted by $f(\theta)$, which is always nonnegative, and then integrated across θ. The first integral thus is negative and the second positive. Whether the *sum* of the two integrals is positive or negative can be seen to depend entirely on the interplay between the IRFs and the θ distribution. A positive sum means that $P_i > P_j$, as in Figure 6.2, and a negative sum means that $P_i < P_j$.

Insight about the meaning of Equation 6.2 may help to understand the methods for investigating an IIO, which are discussed next. Note that if the IRFs have q intersection points, a formula corresponding to Equation 6.2 contains a sum of $q + 1$ integrals: one for the θ interval to the left of the first intersection point, then $q - 1$ for the intervals of θ between two adjacent intersection points, and finally one for the θ interval to the right of the qth intersection point. Within each subsequent interval, the integral's sign is opposite the sign of the integral in the previous interval. The sum of all the positive and negative integrals determines whether $P_i > P_j$ or $P_i < P_j$.

Methods for Investigating IIO

Restscore Method

In this section, we discuss a method that estimates the IRFs of two items by means of the item-rest regression (see Chapter 3), and then compares these estimates to see whether they intersect. This is a direct way of investigating whether two items have an IIO. This is done for all item pairs separately, and the results of these comparisons are combined into one final conclusion about IIO for a set of k items. We discuss the method for two items, and then give a real data example.

Under the null hypothesis of nonintersection, we assume that

$$P_i(\theta) \leq P_j(\theta), \text{ for all } \theta. \quad (6.3)$$

In Chapter 3, we saw that the IRF of item i can be estimated by using the regression of the item score X_i on the restscore $R_{(i)}$; that is, the total score on $k - 1$ items without the item under study, which is item i (Equation 3.1). From a theoretical result (Rosenbaum, 1987a), which we do not discuss in detail here, it follows that when two item-rest regressions are compared, say, for items i and j, the conditioning must be on a summary score that contains neither X_i nor X_j. An obvious choice is the total score on the other $k - 2$ items, defined as

$$R_{(ij)} = \sum_{h \neq i, j} X_h. \quad (6.4)$$

For items i and j, we may now study the inequality

$$P(X_i = 1|R_{(ij)} = r) \leq P(X_j = 1|R_{(ij)} = r), \text{ for } r = 0, \dots, k - 2, \quad (6.5)$$

which estimates the theoretical inequality in Equation 6.3. Other summary scores could be used for conditioning (Rosenbaum, 1987a), but the restscore $R_{(ij)}$ is a convenient choice because it is an unweighted sum score based on $k - 2$ items, comparable to the total score X_+ based on k items and, therefore, like X_+ it stochastically orders θ; see Equation 2.9, and also refer to Exercise 2.2 in Chapter 2.

Equation 6.5 can be used to study the IIO property for pairs of items. The probabilities in Equation 6.5 can be estimated from the sample as fractions of the group with restscore $R_{(ij)} = r$ that have a score of 1 on the studied item. Suppose that in the *total* sample it is found that $P_i < P_j$; this ordering then can be taken as the expected ordering for the fractions in each of the restscore groups. This means that an ordering of fractions in a particular restscore group that contradicts this expected ordering provides evidence that the IRFs of items i and j intersect. Going through all pairs of items and combining all results means investigating IIO for all k items.

An Empirical Example: Transitive Reasoning

The restscore method is incorporated in the computer program MSP (Molenaar & Sijtsma, 2000). We used MSP to analyze the data obtained with 10 transitive reasoning tasks described earlier in Chapter 3 (also see Table 3.1). The two pseudo tasks were left out of the analysis. To demonstrate the use of the item-rest regression method for investigating the possible intersection of two IRFs, we look in detail at Items 4 and 3. Item 4 (weight; 4 objects; equalities) was identified earlier as *not* fitting in with seven other items including Item 3 (weight; 3 objects; inequalities) that together formed a scale.

In the last two columns, Table 6.1 shows the sample fractions that are used in Equation 6.5. It may be noted that the three lowest groups contain too few observations for accurate conclusions about the item ordering. A sensible strategy, therefore, is to combine these three groups with the fourth group, thus obtaining an acceptable sample size of 41 observations for the combined group. It may be noted, in general, that by combining adjacent groups, the probabilities from Equation 6.5 are estimated with greater accuracy, but because restscore groups are combined, fewer points of the item-rest regression are estimated. Thus, combining restscore groups has the effect of creating a clearer view of the IRF, but through narrower windows. Using more but smaller groups means a hazier view through wider windows. It may be noted, for example, that including the restscore group with $R_{(ij)} = 5$ to the lowest groups already combined would lead to a combined group with 126 observations, but only four restscore groups would remain for analysis. The trade-off between accuracy and number of points estimated is less of a problem the larger the sample, in combination with a larger number of items and a large variance of the total score X_+.

In the combined group containing 41 observations, total frequencies per pattern of scores on Item 4 and Item 3 (x_4, x_3) equal: 5 (00), 7 (01), 16 (10), and 13 (11) (obtained by summing the appropriate column frequencies; see the boldface line in Table 6.1). It follows that $P[X_4 = 1|0 \le R_{(4,3)} \le 4] =$ $(16 + 13)/41 \approx 0.71$, and $P[X_3 = 1|0 \le R_{(4,3)} \le 4] = (7 + 13)/41 \approx 0.49$. Subtracting these probabilities yields $vi = 0.22$ (vi meaning size of violation) with $Z = 1.68$ (based on an accurate standard normal approximation to the binomial distribution proposed by Molenaar [1970, chap. 3, Formula 5.5], and provided by MSP [Molenaar & Sijtsma, 2000]), which is significant at the one-sided 5% level ($Z_{crit} = 1.645$).

A complete analysis involving all item pairs for minimum group size of 40 revealed two violations: the one involving Items 4 and 3, just discussed, and another involving Items 4 and 7 ($vi = 0.25$ in the group with $0 \le R_{(4,7)} \le 4$ and group size equal to 40, and $Z = 2.04$). Given the large number of opportunities for intersections of item-rest

Table 6.1   Restscore Method Applied to Two Transitive Reasoning Tasks.

Groups Based on $R_{(4,3)}$			Frequencies per Score Pattern						$P[X_i = 1|R_{(4,3)}]$	
Group	Lo-Hi	n	00	01	10	11	vi	Z	i = 4	i = 3
1	0–1	2	1	0	1	0	.50	—*	.50	.00
2	2–2	4	2	0	2	0	.50	.69	.50	.00
3	3–3	8	1	1	4	2	.38	.89	.75	.38
4	4–4	27	1	6	9	11	.11	.51	.74	.63
1–4	**0–4**	**41**	**5**	**7**	**16**	**13**	**.22**	**1.68**	**.71**	**.49**
5	5–5	75	0	18	13	44			.76	.83
6	6–6	127	1	23	7	96			.81	.94
7	7–7	117	2	23	4	88			.79	.95
8	8–8	65	0	13	1	51			.80	.98
Total		425	8	84	41	292			.78	.88

*No calculations due to low frequencies.

regressions, we do not consider two significant reversals as seriously menacing to the assumption of nonintersection of the 10 IRFs.

Variations on the Restscore Method

Two other methods for comparing pairs of IRFs exist. These methods are variations of the restscore method just discussed. Actually, both methods are based on checking inequalities of the kind

$$P(X_i = 1|S_{(ij)} = s) \leq P(X_j = 1|S_{(ij)} = s), \text{ for each } S_{(ij)} = s. \quad (6.6)$$

In Equation 6.6, $S_{(ij)}$ is a summary of one or more item scores, with the exception of X_i and X_j. The restscore method was defined as $S_{(ij)} = R_{(ij)}$, the total score on $k - 2$ items, not including items i and j; see Equation 6.5.

The first alternative method to be discussed (next) assumes that

$$S_{(ij)} = X_h, h \neq i, j; \quad (6.7)$$

that is, the conditioning simply is on a single item score X_h other than X_i and X_j. The variation on the restscore method that uses Equation 6.7 is denoted the *item-splitting method*, because the ordering of the probabilities is checked in two subgroups based on splitting the total group into one group with item score $X_h = 0$ and another group with $X_h = 1$.

The second alternative method assumes that

$$S_{(ij)} \in s_{(ij)}, \text{ with } s_{(ij)} = \{0, \ldots, r\}, \{r + 1, \ldots, k - 2\};$$
$$r = 0, 1, \ldots, k - 3, \tag{6.8}$$

or, equivalently, $S_{(ij)}$ corresponds to

$$R_{(ij)} \leq r \text{ and } R_{(ij)} > r; \; r = 0, 1, \ldots, k - 3, \tag{6.9}$$

where $R_{(ij)}$ is the restscore, defined in Equation 6.4, and also used as the conditioning variable in the restscore method. The value of $S_{(ij)}$ is substituted in Equation 6.6. The variation of the restscore method that uses Equation 6.8 (or Equation 6.9) is called the *restscore-splitting method*, because the conditioning on a fixed restscore value splits the total group into two subgroups, which is repeated for all meaningful values of the restscore. The item-splitting method and the restscore-splitting method are discussed next in more detail.

Item-Splitting Method

Substituting X_h for $S_{(ij)}$ in Equation 6.6, and explicitly distinguishing $X_h = 1$ and $X_h = 0$, yields

$$P(X_i = 1|X_h = 1) \leq P(X_j = 1|X_h = 1); \tag{6.10a}$$

and

$$P(X_i = 1|X_h = 0) \leq P(X_j = 1|X_h = 0), \tag{6.10b}$$

respectively. These inequalities demonstrate that the ordering of items i and j is the same in the group with a score of 1 on item h and the group with a score of 0 on that item.

The *restscore* method (Equation 6.5) and the *item-splitting* method (Equations 6.10a and 6.10b) have in common that each studies the ordering of a pair of items across subgroups based on an observable statistic that stochastically orders θ (Chapter 2). For the *restscore* method, this means the following: The restscore $R_{(ij)}$ divides the total group into $k - 1$ disjoint subgroups. The next restscore group is characterized by a θ distri-

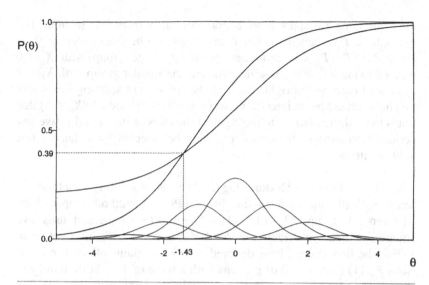

Figure 6.3   Two Intersecting IRFs ($\theta^* = -1.43$), and Normal-Shaped Conditional Distributions of θ Given Restscore $R_{(ij)}$.

bution that overlaps the previous θ distribution but lies farther to the right and has a mean, $E[\theta|R_{(ij)}]$, that is higher than that of the previous restscore group. Thus by letting $R_{(ij)}$ increase from 0 to $k-2$, we obtain information about the ordering of item i and item j within subgroups that gradually shift to the right on the θ scale; see Figure 6.3. In small to moderate samples, the groups based on low and high restscores will often contain few observations and thus provide little power to detect reversals of the expected item ordering. Combining adjacent restscore groups, as illustrated in the previous empirical data example, may make this problem less serious, but at the cost of fewer groups available for studying IIO.

The *item-splitting* method essentially does the same as the restscore method but uses only two disjoint subgroups: one with $X_h = 0$ and the other with $X_h = 1$. Provided the proportion-correct P_h is not very low or very high, these subgroups may be sufficiently large to provide high power to detect different item orderings if the intersection is somewhere between, roughly, $E[\theta|X_h = 0]$ and $E[\theta|X_h = 1]$. If the intersection occurs outside this interval, it may be obscured by the particular grouping using X_h (also see Exercise 6.5 in this chapter). Because we use all items h ($h \neq i, j$) to compare item i and item j by means of Equations 6.10a and 6.10b, this will be less of a problem, in particular if the P_h values of these

$k - 2$ grouping variables show a good variation from low to high. For example, if P_1 is much smaller than 0.5 and we have an ordering $P_1 < P_2 < \ldots < P_k$, P_1 will produce a relatively large group with $X_1 = 0$ located to the left of the θ scale, whereas the smaller group with $X_1 = 1$ is located more to the right. Each of the subsequent splitting items will be characterized by an interval between $E[\theta|X_h = 0]$ and $E[\theta|X_h = 1]$ that tends to be shifted farther to the right. If the IRFs of item i and j have any serious intersections, these will probably be detected by at least a few splitting items.

The P(++)/P(−−) Methodology. The item-splitting method is mathematically identical to the **P(++)/P(−−)** method proposed by Mokken (1971, pp. 132–133). The **P(++)/P(−−)** method uses two square symmetric matrices of order $k \times k$ to investigate intersection of IRFs. The first matrix, here denoted **P(++)**, contains all joint proportions $P_{ij}(11)$ (with $i \neq j$) of persons with a score of 1 on both item i and item j. The second matrix, **P(−−)**, contains all joint proportions $P_{ij}(00)$ (with $i \neq j$) of persons having 0 scores on both items. We do not consider proportions $P_{ii}(11)$ or $P_{ii}(00)$ because their estimation would require scores from the same persons obtained on independent replications of the items. Obviously, we have results from only one test administration available.

Given that $P_i(\theta) \leq P_j(\theta)$ for each θ, if we order rows and columns of **P(++)** and **P(−−)** by increasing P_i value, it can be shown that in each row h $(h = 1, \ldots, k; h \neq i, j)$ of these matrices,

$$P_{hi}(11) \leq P_{hj}(11); \qquad (6.11a)$$

and

$$P_{hi}(00) \geq P_{hj}(00), \qquad (6.11b)$$

respectively. This means that the rows, and by symmetry the columns, of the **P(++)** matrix are nondecreasing, and of **P(−−)** nonincreasing. The proportions in Equations 6.11a and 6.11b can be estimated as sample fractions from empirical data. Violations of the expected orderings are in conflict with an IIO.

It can be shown that Equations 6.11a and 6.11b do not imply each other, but provide independent means for investigating an IIO. Similar inequalities with respect to joint proportions involving (10) or (01)

Table 6.2 $P(++)$ (Upper Panel) and $P(--)$ (Lower Panel) Matrices for 10 Transitive Reasoning Tasks.

Item	9	10	4	5	2	7	3	1	8	6
P_i	.30	.52	.78	.80	.81	.84	.88	.94	.97	.97
9 .30		.22	.23	.25	.24	.28	.28	.29	.30	.30
10 .52	.22		.39	.44	.42	.48	.49	.51	.51	.52
4 .78	.23	.39		.64	.69	.67	.69	.73	.76	.76
5 .80	.25	.44	.64		.66	.69	.73	.77	.78	.79
2 .81	.24	.42	.69	.66		.68	.72	.76	.78	.80
7 .84	.28	.48	.67	.69	.68		.79	.82	.83	.84
3 .88	.28	.49	.69	.73	.72	.79		.86	.88	.88
1 .94	.29	.51	.73	.77	.76	.82	.86		.92	.93
8 .97	.30	.51	.76	.78	.78	.83	.88	.92		.96
6 .97	.30	.52	.76	.79	.80	.84	.88	.93	.96	
9 .30		.40	.14	.15	.13	.13	.10	.05	.03	.02
10 .52	.40		**.08**	.12	.09	.12	**.09**	.04	.03	.02
4 .78	.14	.08		.05	.09	.04	.02	.01	.01	.01
5 .80	.15	.12	.05		.04	.04	.04	.02	.01	.01
2 .81	.13	.09	.09	.04		.03	.02	.01	.01	.01
7 .84	.13	.12	**.04**	.04	.03		**.06**	.03	.02	.02
3 .88	.10	.09	.02	.04	.02	.06		.04	.02	.02
1 .94	.05	.04	**.01**	.02	.01	.03	**.04**		.01	.01
8 .97	.03	.03	**.01**	.01	.01	.02	**.02**	.01		.01
6 .97	.02	.02	**.01**	.01	.01	.02	**.02**	.01	.01	

Note: Violations of expected ordering of items 4 and 3 are printed in boldface.

patterns are implied by Equations 6.5a and 6.5b, however, and such inequalities thus are redundant for investigating an IIO (Sijtsma & Junker, 1996). Finally, Equations 6.11a and 6.11b can be transformed easily into Equations 6.10a and 6.10b, respectively, and the other way around. This shows that the item-splitting method and the $P(++)/P(--)$ method are equivalent methods for investigating IIO.

An Empirical Example: Transitive Reasoning

Although the method considers all items, here we use only Items 4 and 3 for illustration purposes. Their IRFs were compared using the $P(++)$ and $P(--)$ matrices (Table 6.2); this is the method incorporated

Table 6.3   Item-Splitting Method: Ordering of Item 4 and Item 3 in the Subgroup With Item h Incorrect and the Subgroup With Item h Correct.

Splitting Item h	P_h	$X_h = 0$		$X_h = 1$	
		Item 4	*Item 3*	*Item 4*	*Item 3*
9	.30	.79	.86	.76	.95
10	.52	**.83**	**.81**	.74	.95
5	.80	.74	.80	.79	.91
2	.81	.51	.89	.85	.88
7	.84	**.76**	**.61**	.79	.94
1	.94	**.84**	**.40**	.78	.91
8	.97	**.79**	**.29**	.78	.90
6	.97	**.73**	**.36**	.78	.90

Note: Violations of expected ordering are printed in boldface.

in MSP. Comparison of the columns for Item 4 and for Item 3 [third and seventh columns in the $P(++)$ matrix] reveals that the expected ordering of joint proportions is found for each of the other eight items (rows). In the $P(--)$ matrix, the expected ordering is violated for row Items 10, 7, 1, 8, and 6. For row Item 10, $vi = 0.01$ with $Z = 0.25$ (see Molenaar & Sijtsma, 2000, for an explanation of the statistical test); for row Item 7, $vi = 0.02$ with $Z = 1.60$; for row Item 1, $vi = 0.03$ with $Z = 2.93$; for row Item 8, $vi = 0.02$ with $Z = 2.07$; and for row Item 6, $vi = 0.01$ with $Z = 1.23$. Further analysis of the $P(++)$ and $P(--)$ matrices revealed that the items 2, 4, 7, 3, and 5 were involved in 6, 9, 6, 5, and 4 significant violations, respectively, each of which occurred in the $P(--)$ matrix.

Because all the significant violations were revealed by the $P(--)$ matrix, the intersections occurred at the lower end of the scale. This was corroborated by the inspection of the inequalities in Equations 6.10a and 6.10b for Item 4 and Item 3 (Table 6.3). The expected item ordering was violated for five splitting items, four of which had high P_h values and thus small groups with $X_h = 0$, which are located at the relatively low end of the scale.

Restscore-Splitting Method

Using Equation 6.9 and substituting for $S_{(ij)}$ in Equation 6.6 yields

$$P(X_i = 1|R_{(ij)} \le r) \le P(X_j = 1|R_{(ij)} \le r) \tag{6.12a}$$

and

$$P(X_i = 1|R_{(ij)} > r) \le P(X_j = 1|R_{(ij)} > r), \tag{6.12b}$$

with $r = 0, \ldots, k - 3$. For a fixed value r, the restscore-splitting method divides the total group into two subgroups, one with a relatively low restscore and the other with a relatively high restscore. The higher-restscore group has a higher mean θ due to the stochastic ordering property. By successively varying the r value from low to high and checking the inequalities in Equation 6.12a and 6.12b for each r, the IRFs of items i and j are investigated for intersections at increasingly higher regions along the θ scale.

Restscore-splitting combines the merits of the *restscore* method and the *item-splitting* method. Because only two subgroups are formed for each r, the power to detect any intersections is high, comparable with that of the item-splitting method. By conditioning on the restscore based on $k - 2$ items rather than a single item h, a more reliable subdivision into subgroups is realized, comparable with that of the restscore method.

An Empirical Example: Transitive Reasoning

The restscore-splitting method is implemented in the MSP computer program (Molenaar & Sijtsma, 2000). For Items 4 and 3, Table 6.4 gives the fractions that estimate the probabilities in Equations 6.12a and 6.12b, for five subsequent dichotomizations of the restscore. Dichotomizations for $r \le 2$ have not been tabulated due to small subgroup sizes. The restscore-splitting method shows that the IRFs of Items 4 and 3 intersect at the lower end of the scale (one-sided Z values significant at the 5% level).

A Summary Measure for IIO of k Items

The restscore method, the item-splitting method (identical to the $P(++)/P(--)$ methodology), and the restscore-splitting method are diagnostics for singling out individual items whose IRFs cross with the IRFs of other items from the same test. Prior to using these diagnostics, a more general impression about the degree to which intersections occur among k IRFs may be desirable. Coefficient H^T (Sijtsma & Meijer, 1992) may be used for this purpose. H^T compares the score patterns of 0s and 1s on k items produced by n individuals. The more similar these item score patterns, the

Table 6.4   Restscore-Splitting Results for Two Transitive Reasoning Tasks.

Dich	$R_{(ij)}$	n	$P(X_4 = 1)$	$P(X_3 = 1)$	Z
1	0–3	14	**.64**	**.21**	1.81
	4–8	411	.79	.91	
2	0–4	41	**.71**	**.49**	1.68
	5–8	384	.79	.93	
3	0–5	116	**.74**	**.71**	.41
	6–8	309	.80	.95	
4	0–6	243	.78	.83	
	7–8	182	.79	.96	
5	0–7	360	.78	.87	
	8	65	.80	.98	

Note: Violations of expected ordering are printed in boldface.

higher the H^T. Similarly, a coefficient H^T_a can be defined for one individual with respect to the $n - 1$ other individuals. The more the score pattern of Person a resembles the average score pattern produced by the other $n - 1$ respondents, the higher the H^T_a.

It has been shown (Sijtsma & Meijer, 1992) that H^T is informative about the intersection of IRFs. The theoretical justification for this result is beyond the scope of this book. We provide only the following result: For a set of k items having nonintersecting IRFs,

$$0 \le H^T \le 1, \tag{6.13}$$

where $H^T = 0$ if and only if all k IRFs coincide, with the possible exception of at most one θ value. The practical meaning of this result is that the higher a positive H^T, the more likely it is that the IRFs do not intersect. Thus, if H^T has a high positive value, we expect few intersections of IRFs.

Although a positive H^T is *indicative* of nonintersection, *specific* positive H^T values do not tell the researcher *exactly* how many intersections have occurred or how many items are involved. For example, even $H^T = 0.20$ can be indicative of k IRFs that are close together without intersecting, and also of IRFs that are farther apart but intersect at some points on the θ scale. Because the assumption that IRFs do not intersect implies that H^T is positive (Equation 6.13 is based on this assumption), it certainly is true that negative H^T values mean that there *are* intersections. To reduce the uncertainty about the meaning of *positive* H^T values, the H^T_a

coefficient can be used in addition to H^T in the following way (Sijtsma & Meijer, 1992):

> If $H^T \geq 0.3$, and less than 10% of the persons have negative H^T_a values, then it is assumed for all practical purposes that the k IRFs do not intersect.

This is a rule of thumb. If it is satisfied, this does not automatically imply that there are no intersections. However, simulations have provided some evidence that this rule works rather well in practice (Sijtsma & Meijer, 1992).

For the transitive reasoning data (Chapter 3) without the two pseudo tasks, $H^T = 0.47$, and the percentage of negative H^T_a values was 12. Because the latter result does not satisfy the rule of thumb, this may be sufficient reason for the researcher to use the restscore method, the item-splitting method [the $\mathbf{P}(++)/\mathbf{P}(--)$ methodology], or the restscore-splitting method to find the items that are most often involved in intersections. These items may be removed from the test in order to obtain an IIO for the remaining items. After removal of Items 2 and 4 (equalities only; see Table 3.1; also see Sijtsma & Junker, 1997, for a justification), H^T increases to 0.65 and the percentage of negative H^T_a values drops to 5.2. The rule of thumb thus is satisfied, and the remaining items are assumed to have an IIO. Results are even better if Item 5 (the only task with an ordering opposite to the ordering of the other tasks; see Table 3.1; also see Sijtsma & Junker, 1997) is removed: $H^T = 0.75$, and the percentage of negative H^T_a is now 3.0.

Reliability of the Total Score and of Item Scores

The classical way to determine measurement accuracy is to estimate the degree to which the test score X_+ can be repeated under a hypothetical, independent repetition of the test. The degree of repeatability is the *reliability* of a test score. In practice, independent repetitions are not available, and a rank correlation or a product-moment correlation between X_+ and a replicated total score X'_+ thus cannot be computed. For reasons of consistency, in a nonparametric context, a rank correlation would be most appropriate. We use the product-moment correlation, however, because it allows us to estimate a coefficient of measurement accuracy based on only one observable test score. Moreover, a drawback of the rank correlation is that we would be deprived of a standard measurement error, which is very useful in individual diagnosis or when individuals' scores are compared.

The product-moment correlation between two hypothetical replications X_+ and X'_+, denoted $\rho_{X_+X'_+}$, which is the classical reliability coefficient, is therefore used. Mokken (1971, pp. 142–147) proposed two related methods for estimating the classical reliability of X_+ that are based on the nonintersection of the IRFs of the items for which X_+ is calculated. A third and related method (denoted method MS) was proposed by Sijtsma and Molenaar (1987). All three methods estimate the *unobservable* proportions $P_{ii}(11)$ $(i = 1, \ldots, k)$, which are the diagonal elements of the $\mathbf{P}(++)$ matrix defined in Equation 6.11a, and that stand for the proportion of respondents who have a score of 1 on two independent replications of item i. For example, one method estimates $P_{ii}(11)$ by interpolation of the *observable* proportions $P_{i-1,i}(11)$ and $P_{i,i+1}(11)$, thereby making use of the nondecreasingness of the columns in the $\mathbf{P}(++)$ matrix (this property follows from the nonintersection of the IRFs). For example, in Table 6.2, proportion $P_{33}(11)$, which is at the crossing of row 3 and column 3 (which correspond to Item 4 in the; original item numbering), is estimated using the two neighboring bivariate proportions 0.39 and 0.64. The actual estimation of $P_{ii}(11)$ also involves the popularity of the items $i-1$, i, and $i+1$ (Items 10, 4, and 5, using original item numbers); and this estimation is repeated for all k items. The results are then used to estimate the reliability of X_+, but outlining the whole procedure in full detail would go beyond the level of detail of this book.

Sijtsma and Molenaar (1987) showed by means of simulated data that these three methods had smaller bias with respect to $\rho_{X_+ X'_+}$ than classical methods such as Cronbach's (1951) alpha coefficient and Guttman's (1945) lambda-2 coefficient. Furthermore, the classical methods had smaller sampling error than the three methods according to Mokken, and to Sijtsma and Molenaar, although differences were small and probably unimportant for practical purposes.

The following example shows how the MS reliability method is used. For X_+ based on 10 transitivity tasks, MSP calculated a method MS reliability of 0.55; after deletion of Items 2 and 4 (see Chapter 3), this method yielded 0.65; and after removal of Item 5 (see Chapter 3) the reliability further increased to 0.68. Now we take a closer look at the three items that were removed. The reader may recall from Chapter 3 that the removed Items 2, 4, and 5 each had low item scalability H_i values (Table 3.2), which is indicative of IRFs with relatively flat slopes. These items thus had a weak relation with the latent trait. In general, such items do not contribute to the reliability of the total score, and sometimes (like here) they may even lower it. Moreover, the presence of these items in the item set obstructs an IIO.

Discussion

The DMM implies the invariant ordering of items in addition to the stochastic ordering of the latent trait by means of the total score. These two ordering properties are powerful tools for many measurement purposes. If the sole purpose of a test application is the ordinal measurement of persons, however, the monotone homogeneity model suffices. This model allows IRFs of all forms as long as they are monotonely nondecreasing. This means that the IRFs are allowed to intersect.

If applications of the test are envisioned that require an equal item ordering for all values of θ, or within different relevant subgroups taken from the population of interest, a model like the DMM is needed: It assumes nonintersecting IRFs and, thus, implies an IIO. In practice, nonintersection of IRFs is a restrictive condition for data to meet, but the result of a fitting DMM is that the IIO property allows the testing of psychological theories by means of item response models, the use of uniform starting and stopping rules in intelligence testing, the unequivocal interpretation of item score patterns when examinees may have produced aberrant item scores, and the investigation of differential item functioning using item ordering in reference and focal groups.

Additional Reading

The DMM has been treated by Mokken (1971, 1997), Mokken and Lewis (1982), and Sijtsma (1998). The concept of invariant item ordering has been discussed by Sijtsma and Junker (1996). These authors also surveyed methods for investigating IIO in real data and gave empirical data examples of the application of each of the methods (also, see Sijtsma & Junker, 1997). The restscore method was discussed in great detail by Rosenbaum (1987a), the item-splitting method by Sijtsma and Junker (1996), and the equivalent $P(++)/P(--)$ method by Mokken (1971; also, see Rosenbaum, 1987b), and the restscore-splitting method by Sijtsma and Junker (1996). Croon (1991) and Hoijtink and Molenaar (1997) studied the DMM from the perspective of latent class analysis.

Excellent methods for investigating differential item functioning have been proposed (see Holland & Wainer, 1993, for an overview) in the context of parametric IRT, and by Shealy and Stout (1993) and Molenaar and Sijtsma (2000) in the context of nonparametric IRT. Person-fit methods have been discussed in the nonparametric context by Meijer and Sijtsma (2001), and Emons, Meijer, and Sijtsma (in press).

Exercises for Chapter 6

6.1. Mention at least two reasons why it could be useful to require that item response functions do not intersect.

6.2. Use Table 6.1 for this exercise. Suppose we want each of the restscore groups to contain at least 100 observations.
   a. Produce a table similar to Table 6.1 for this situation.
   b. One column cannot be produced on the basis of Table 6.1 alone. Which one is that?
   c. What do you conclude about intersection of the IRFs of Items 4 and 3 based on this new table? What would be your conclusion if you knew that $Z = 0.41$ (produced by the MSP program)?
   d. Compare the result from (c) with the result based on Table 6.1 discussed in the text.
   e. Does your new table provide information about monotonicity? If so, what conclusion can be drawn?

6.3. a. Show that Equation 6.11a for the $P(++)$ matrix implies Equation 6.10a for the item-splitting method.
   b. Analogously, show that Equation 6.11b for the $P(--)$ matrix implies Equation 6.10b for the item-splitting method.

6.4. Violations found in the $P(++)$ and $P(--)$ matrices (Equations 6.11a and 6.11b) tend to be smaller than violations based on the item-splitting method (Equations 6.10a and 6.10b). Does this mean that we should take the information provided by Equations 6.11a and 6.11b less seriously than that provided by Equations 6.10a and 6.10b?

6.5. In the following numerical example (Sijtsma & Junker, 1996), determine whether or not the item-splitting method (Equations 6.10a and 6.10b) reveals that the IRFs of items i and j intersect. Table 6.5 shows a discrete latent trait θ with four values and a uniform distribution. The splitting item h is answered incorrectly for $\theta \leq 2$, and correctly otherwise. The conditional distributions given the score on the splitting item are given in the fourth and fifth columns. The last two columns show the four points of both IRFs that are meaningful for this discrete θ.

Table 6.5

θ	$P(\theta)$	X_h	$P(\theta\|X_h = 0)$	$P(\theta\|X_h = 1)$	$P_i(\theta)$	$P_j(\theta)$
1	.25	0	.5	.0	.1	.2
2	.25	0	.5	.0	.3	.4
3	.25	1	.0	.5	.5	.7
4	.25	1	.0	.5	.9	.8

a. What can be said about the relationship between the latent trait and the splitting item? Is this a realistic situation?

b. How is $P(\theta = 1|X_h = 0)$ (fourth column, first entry) obtained? Similarly, how is $P(\theta = 1|X_h = 1)$ (fifth column, first entry) obtained?

c. Do the IRFs of items i and j intersect? How do you know?

d. Show that the item-splitting method does not reveal the result of (c).

Answers to Exercises for Chapter 6

6.1. When a theory predicts that some items are more difficult than others, this ordering should be found for all latent trait values. This is also desirable when starting and stopping rules are used. Moreover, an invariant item ordering facilitates the study of aberrant response patterns and the study of differential item functioning.

6.2. a. Results for item-rest regressions with minimum group size equal to 100 are in Table 6.6:

b. The column that holds the information about the statistic Z. To produce that statistic, we would have to know how it is calculated.

c. The lowest group still shows a reversal of the expected ordering. Given that $Z = 0.41$, $vi = 0.03$ is not significant. No evidence of reversal.

Table 6.6

Groups Based on $R_{(4,3)}$			Frequencies per Score Pattern						$P[X_i = 1|R_{(4,3)}]$	
Group	Lo-Hi	n	00	01	10	11	vi	Z	$i = 4$	$i = 3$
1	0–5	116	5	25	29	57	.03	?	.74	.71
2	6–6	127	1	23	7	96			.81	.94
3	7–8	182	2	36	5	139			.79	.96
Total		425	8	84	41	292	.03		.78	.88

    d. For a minimum group size of 1, in the four lowest restscore groups, four nonsignificant reversals were found; for the four lowest groups combined ($n = 41$), however, the reversal found was significant. For minimum group size of 100, the reversal found was not significant. This seems to be due to the addition of a large group of 75 respondents in which there was no reversal. This group obscures the reversal in the lowest four groups. Because the reversal is consistent in the four lowest groups, and significant if these groups are joined, our advice is to take it seriously.

    e. The last two columns provide information about monotonicity for Items 4 and 3, respectively. For Item 4, there is one violation with $vi = 0.02$.

6.3. a. Note that $P_{hi}(11) = P(X_i = 1|X_h = 1) \times P(X_h = 1)$, and that $P_{hj}(11) = P(X_j = 1|X_h = 1) \times P(X_h = 1)$. Thus, it follows immediately that if $P_{hi}(11) \leq P_{hj}(11)$ then, $P(X_i = 1|X_h = 1) \leq P(X_j = 1|X_h = 1)$.

    b. Use $P_{hi}(00) = P(X_i = 0|X_h = 0) \times P(X_h = 0)$, and $P_{hj}(00) = P(X_j = 0|X_h = 0) \times P(X_h = 0)$. Because $P_{hi}(00) \geq P_{hj}(00)$, we have $P(X_i = 0|X_h = 0) \geq P(X_j = 0|X_h = 0)$. Considering $X_i = 1$ rather than $X_i = 0$, it immediately follows that $P(X_i = 1|X_h = 0) \leq P(X_j = 1|X_h = 0)$.

6.4. No. To see this you have to note that each fraction used to estimate the proportions in Equations 6.11a and 6.11b is based on the complete sample. Each of the fractions used to estimate the probabilities in Equations 6.10a and 6.10b is based on part of the sample, however; that is, the part with $X_h = 1$ (Equation 6.10a) or $X_h = 0$ (Equation 6.10b). Consequently, the fractions used in Equations 6.11a and 6.11b are estimated more accurately. This compensates to some extent for the smaller differences found.

6.5. a. The rank correlation between θ and X_h is high. In practice, such correlations are often positive, but not as high as it is here.

    b. Respondents with $X_h = 0$ have $\theta = 1$ or 2 with equal probability; hence, $P(\theta = 1|X_h = 0) = 0.5$. Respondents with $X_h = 1$ do not have $\theta = 1$; hence, $P(\theta = 1|X_h = 1) = 0.0$.

    c. Simply read Table 6.5: For $\theta = 1, 2$, and 3, $P_i(\theta) < P_j(\theta)$, but for $\theta = 4$, $P_i(\theta) > P_j(\theta)$. The IRFs thus intersect between $\theta = 3$ and $\theta = 4$.

    d. Equation 6.10a: Within the group with $X_h = 1$, respondents have either $\theta = 3$ or 4; both values have probability 0.5. Thus the probability of having item i correct (left-hand side in Equation 6.10a) is found by weighting $P_i(\theta = 3)$ and $P_i(\theta = 4)$ by 0.5 each; this yields $0.5 \times (0.5 + 0.9) = 0.7$. Similarly, we have $P(X_j = 1|X_h = 1) = 0.5 \times (0.7 + 0.8) = 0.75$.

    Equation 6.10b: $P(X_i = 1|X_h = 0) = 0.5 \times (0.1 + 0.3) = 0.2$; and $P(X_j = 1|X_h = 0) = 0.5 \times (0.2 + 0.4) = 0.3$.

Both Equations 6.10a and 6.10b are satisfied, but this does not reveal the intersection of the IRFs between $\theta = 3$ and $\theta = 4$. In this hypothetical example, we have only three items. Thus, only item h can be used to investigate the possible intersection of the IRFs of items i and j using the item-splitting method. In realistic tests, the number of items k usually is much higher than 3, and $k - 2$ items can each be used as splitting items. This increases the probability of finding intersections of the IRFs of items i and j if there are any.

7

Extension of Nonparametric IRT
to Polytomous Item Scores

So far, we have considered NIRT models for dichotomous items. Regularly, however, respondents are offered items with several ordered answer categories. For example, an item measuring attitude toward abortion may be presented as, "I think a woman should be able to decide on her own whether she wants to have an abortion," with ordered answer categories *strongly agree, agree, neutral, disagree*, and *strongly disagree*. An item measuring extroversion by means of self-rating might be: "When I am in the company of others, I give my opinion when I feel the need to do so," with ordered answer categories *almost never, seldom, often*, and *almost always*. Achievement items also can have ordered answer categories, but here the answer categories are not visible to the respondent and scoring takes place afterward. For example, solutions to examination items may be rated *fully incorrect, partially correct*, and *fully correct*. This may require a written account by the respondent on how he or she solved the problem, and based on this, the tester may decide that the account shows enough evidence of at least some understanding of the problem and its solution. Consequently, one credit point (*partially correct*) instead of zero (*fully incorrect*) could be assigned, and a fully correct solution could be scored with two credit points.

The examples make clear that the number of answer categories may vary among tests or questionnaires. In this chapter, we assume that all items used in a test or questionnaire to measure a particular latent trait have the same number of answer categories. Although we could model items with different numbers of categories, each of these items measuring the same θ, we ignore this possibility because (a) *in practice*, items in a questionnaire measuring the same θ usually have the same number of answer categories, and (b) often there is no *theoretical* justification for using different numbers of answer categories, which would result in the assignment of different weights to the items, simply because with one item, more points could be earned than with another item. A less compelling, albeit not irrelevant, reason is that the notation in this chapter would be more complicated if the number of answer categories were allowed to vary across the k items in a test.

The extensions of the MHM and the DMM that are discussed in the present chapter allow us to handle items with ordered scores, denoted *polytomous items* (first mentioned in Chapter 1). We assume that items are scored such that $X_i = 0, 1, \ldots, m$, for all $i = 1, \ldots, k$. This means that an item has $m + 1$ ordered answer categories. The test score for k polytomous items is defined as

$$X_+ = \sum_{i=1}^{k} X_i, \text{ with } X_+ = 0, 1, \ldots, m \times k. \tag{7.1}$$

For example, if $k = 3$, and the item scores are 1, 0, and 3, then the test score $X_+ = 4$. It may be noted that this definition extends Equation 2.5 (total score based on dichotomous items). Further, as with the scale of the total score for dichotomous items, the scale of the total score for polytomous items starts at 0, irrespective of the number of items in the test. In practice, we often find that item scoring starts at 1, with possible item scores $X_i = 1, 2, \ldots, w$ (in our notation, $w = m + 1$). This so-called Likert scoring results in total scores, $X_+ = k, k + 1, \ldots, w \times k$, however, thus starting at k. Although both scoring rules lead to the same psychometric results, we prefer the scoring in Equation 7.1 because it assigns 0 points to someone who scored k times in the lowest category, which seems to make more sense at an intuitive level of thought.

Finally, it may be noted that the case of *nominal* answer categories, such as the correct answer and three distracters in a multiple-choice item, or the choice of one's favorite from a list of brands, is not covered by the polytomous NIRT extension discussed here. Our models require that the $m + 1$ answer categories have a meaningful order representing increasingly higher levels of the latent trait. This is a crucial *psychological* assumption underlying the assignment of ordered scores to answer categories. When this assumption is not realistic, the analysis of polytomous scores is likely to produce a misfitting model and misfitting individual items. A well-known example occurs when answer categories have been scored in the wrong direction given the wording of the item. For example, this may happen when a questionnaire measuring the attitude toward abortion contains both items expressing a negative attitude and items expressing a positive attitude. The reader is invited to verify that an *agree* on a *negatively* worded item indicates a *low* standing on the attitude favoring abortion, and on a *positively* worded item a *high* standing. In the first case, the item score should be 0, and in the second case *m*; and so on. Ignoring this, thus giving the same score for *agree* (and other answer categories) irrespective of the wording of the item, produces several negative covariances and scalability coefficients. Other examples, where the problem of misfit is more difficult to remedy, may represent an inadequate operationalization of the construct to be measured. Statistical analysis using a polytomous IRT model may provide indications of which items to reject from the test or how to rephrase misfitting items adequately.

Definition of Nonparametric Polytomous IRT

As before, we start with the assumptions of unidimensionality and local independence at the item level; see Chapter 2. As with the MHM for dichotomous items, the MHM for polytomous items puts only inequality restrictions on the response functions. Compared with Chapter 2, the only difference is that the assumption of monotonicity is now applied to the so-called item step response functions (ISRFs). In this and the next chapter, index *h* is used to indicate polytomous item scores (previously, *h* denoted an item); therefore, the ISRF is defined as

$$P_{ih}(\theta) = P(X_i \geq h|\theta), i = 1, \ldots, k; \text{ and } h = 0, \ldots, m. \quad (7.2)$$

Equation 7.2 gives the probability of scoring at least *h* on item *i*. It may be noted that $h = 0$ leads to a probability of 1 for each θ, which is not informative about item functioning. This means that each item with $m + 1$ answer categories has *m* meaningful ISRFs. The MHM assumes that each of these ISRFs is monotone nondecreasing in θ.

It may be noted that within one item, the *m* meaningful ISRFs cannot intersect *by definition*. To see why this is true, we write

$$P_{ih}(\theta) = \sum_{g=h}^{m} P(X_i = g|\theta). \quad (7.3)$$

That is, the probability of having at least a score of h equals the sum of probabilities of having a score of exactly h plus the probability of having a score of exactly $h + 1$, and so on. Similarly, the next ISRF, $P_{i,h+1}(\theta)$, can be written as

$$P_{i,h+1}(\theta) = \sum_{g=h+1}^{m} P(X_i = g|\theta). \tag{7.4}$$

Subtracting $P_{i,h+1}(\theta)$ (Equation 7.4) from $P_{ih}(\theta)$ (Equation 7.3) yields

$$P_{ih}(\theta) - P_{i,h+1}(\theta) = P(X_i = h|\theta). \tag{7.5}$$

In other words, the hth ISRF and the $(h + 1)$st ISRF of item i differ by a nonnegative amount, $P(X_i = h|\theta)$, which may vary across θ, and which causes $P_{ih}(\theta)$ to be at least as high as $P_{i,h+1}(\theta)$ for all θs. The ISRFs may coincide for some θs when $P(X_i = h|\theta) = 0$ for those θs, but the important point here is that within one item, *the m ISRFs cannot intersect*. This is typical of the MHM for polytomous items.

Figure 7.1 displays ISRFs for two items with five ordered answer categories each. It can be seen that the ISRFs of the same item do not

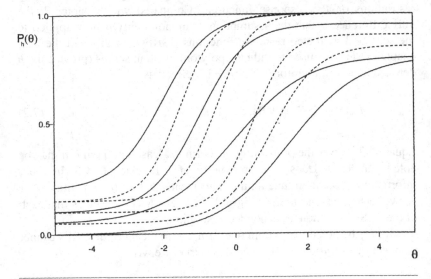

Figure 7.1   Itemstep Response Functions for Two Items With Five Ordered Answer Categories Each.

intersect. However, under the MHM, the ISRFs of different items are allowed to intersect; this is the situation shown in Figure 7.1. In this respect, the MHM for polytomous items is similar to the MHM for dichotomous items, which also allows IRFs of *different* items to intersect.

Ordering Persons

As with the MHM for dichotomous items, measurement by means of the MHM for polytomous items also uses total score X_+ (Equation 7.1) for ordering respondents on θ. The MHM for polytomous items implies the stochastic ordering of total score X_+ by latent trait θ; see Equation 2.11. Hemker, Sijtsma, Molenaar, and Junker (1997) have shown that, unfortunately, when reversing roles, X_+ does not always order respondents on θ in the stochastic ordering sense discussed in Chapter 2 (Equation 2.9). This means that at the *theoretical* level, the ordering on the observable X_+ may give a somewhat distorted view of the ordering on the latent trait θ. Van der Ark (2000) has shown, by using simulated polytomous test data, that at the *practical* level, X_+ may be expected to order respondents *correctly* on θ for many tests and many distributions of θ. Moreover, distortions occur only in rare cases and almost always involve respondents that are located closely on the X_+ scale, say, one or two points apart. Due to random measurement error, such respondents are hard to distinguish anyway, and their score differences, even when statistically significant, have little practical use.

Our conclusion is that under the MHM, we may confidently use X_+ as a proxy for ordering respondents on θ. This is an extremely pleasant result, because it justifies, from a practical point of view, the use of the MHM for ordering persons on the latent trait. As concerns person ordering, Table 7.1 presents the frequency distribution of X_+ based on three polytomous items describing strategies to cope with industrial malodor (Items 1, 2, and 4 in Table 5.2). Each item is scored 0 (*never*), 1 (*rarely*), 2 (*often*), or 3 (*always or almost always*), so that $X_+ = 0, \ldots, 9$. Under the MHM, we may consider X_+ as a proxy for θ and, thus, in Table 7.1 from the 55 persons scoring 0 (response *never* to all three items) up to the 145 persons scoring 9 (response *always* to all three items), one may infer an increasing amount of the latent trait, which could be labeled "coping by means of practical household measures." As concerns item analysis, the MHM is an IRT model that poses relatively weak (in the sense of not highly demanding) restrictions on the data and thus is expected to fit data

Table 7.1   Frequency Distribution of X_+, and Mean, Standard Deviation, Skewness, and Kurtosis Based on Industrial Malodor Items 1, 2, and 4 (Table 5.2), Scored 0, 1, 2, 3.

X_+	0	1	2	3	4	5	6	7	8	9
Frequency	55	81	83	178	56	51	73	65	41	145

Mean	4.57	Skewness	0.21
Stand. Dev.	2.92	Kurtosis	−1.24

sets without the need to reject many items first. In Chapter 8 we give examples of item analyses using real data that illustrate the practical use of the MHM.

Nonnegative Covariances, Item Steps, and Scalability Coefficients

Nonnegative Covariances

It can be shown (Molenaar, 1997) that Theorem 4.1 also holds for the polytomous MHM: That is, all covariances between item pairs are nonnegative. Just like with IRFs for dichotomous items, the condition of nonnegative covariances for item pairs is necessary but not sufficient for nondecreasing ISRFs, and sampling error may cause a covariance estimate to be negative although the population covariance is positive, and vice versa.

Item Steps

In what follows, we need the concept of the *item step*. We define a binary response variable, Y_{ih}, with item index $i = 1, \ldots, k$, and score index $h = 1, \ldots, m$, as

$$Y_{ih} = 0 \text{ if } X_i < h; \tag{7.6a}$$

and

$$Y_{ih} = 1 \text{ if } X_i \geq h. \tag{7.6b}$$

In words: The score on Y_{ih} tells us whether the respondent has at least an item score of h ($Y_{ih} = 1$) or not ($Y_{ih} = 0$). In the example given to illustrate Equation 7.1, for $k = 3$, and item scores 1, 0, and 3, using Equations 7.6a and 7.6b we find that $X_1 = 1$ implies $Y_{11} = 1$ and $Y_{12} = Y_{13} = 0$; $X_2 = 0$ implies $Y_{21} = Y_{22} = Y_{23} = 0$; and $X_3 = 3$ implies $Y_{31} = Y_{32} = Y_{33} = 1$. One might also imagine the following response process: When a respondent is faced with a rating scale item, such as, "When I am in the company of others, I give my opinion when I feel the need to do so," with ordered answer categories *almost never, seldom, often,* and *almost always,* the respondent first reflects whether he or she is enough of an extrovert to take the step from the first answer category to the second, which is higher on the latent trait. If not, then $Y_{i1} = 0$ and $X_i = 0$; if yes, then $Y_{i1} = 1$ and $X_i \geq 1$. In the latter case, our respondent wonders next whether he or she can take the step from the second to the third answer category. If not, then $Y_{i2} = 0$ and $X_i = 1$; if yes, $Y_{i2} = 1$ and $X_i \geq 2$. The final step can either result in $Y_{i3} = 0$ and $X_i = 2$, or $Y_{i3} = 1$ and $X_i = 3$.

From the definition of an item step variable, Y_{ih}, we have two results that we need later on. First, from the definition of Y_{ih} (Equations 7.6a and 7.6b), it follows that

$$P(Y_{ih} = 1|\theta) = P(X_i \geq h|\theta). \tag{7.7}$$

That is, the conditional probability of taking the hth item step of item i is the ISRF. Second, it may be noted that for an item with $m + 1$ answer categories, there are m item step variables, Y_{ih}, $h = 1, \ldots, m$. Now, we may rewrite the total score X_+ as

$$X_+ = \sum_{i=1}^{k} X_i = \sum_{i=1}^{k} \sum_{h=1}^{m} Y_{ih}, \text{ with } X_+ = 0, 1, \ldots, m \times k. \tag{7.8}$$

That is, the total score X_+ counts the total number of item steps successfully taken on k items. We need the item step concept when discussing scalability coefficients for polytomous items.

Scalability Coefficients

The H coefficient for scalability of an item pair is based on the same principles as for dichotomous items, but its definition is a little more complicated. A pair of items has $2m$ item steps that may be jointly ordered from easiest to most difficult. We need to know two things about this ordering.

First, we start by defining the proportions of people taking the hth step of item i, P_{ih}. Similar to proportion P_i (Chapter 2, Equation 2.15), which is the proportion of correct answers on a dichotomous item, we obtain P_{ih} by integrating across the distribution $f(\theta)$ of θ:

$$P_{ih} = \int_\theta P(Y_{ih} = 1|\theta) f(\theta)\, d\theta$$

$$= \int_\theta P_{ih}(\theta) f(\theta)\, d\theta, h = 1, \ldots, m.$$

(7.9)

Because for any single item the m ISRFs cannot intersect *by definition*, for two adjacent item, steps h and $h + 1$, it can easily be seen by using Equation 7.5 for the difference between two adjacent ISRFs, that

$$P_{ih} - P_{i,h+1} = \int_\theta [P_{ih}(\theta) - P_{i,h+1}(\theta)] f(\theta)\, d\theta$$

$$= \int_\theta P(X_i = h|\theta) f(\theta)\, d\theta \geq 0.$$

(7.10)

Thus, *by necessity* we always have $P_{ih} \geq P_{i,h+1}$. In general, the fixed order of the m item steps of item i is

$$P_{i1} \geq P_{i2} \geq \ldots \geq P_{im}.$$ (7.11)

Generalizing this result to two items, i and j, and indexing their item steps by g and h, what we have now is that the simultaneous ordering of the $2m$ item steps from these items is reflected by their ordering using the item step proportions P_{ig} ($g = 1, \ldots, m$) and P_{jh} ($h = 1, \ldots, m$). Also, we know that the m item steps of the *same* item have a fixed order. It follows that only the item steps from *different* items do not have an a priori fixed order. That is, for item i by necessity $P_{i2} \geq P_{i3}$, but for items i and j, either $P_{i3} \geq P_{j2}$ or $P_{i3} < P_{j2}$. In practice, this ordering is determined by the data.

Second, for an ordering of $2m$ item steps, and arbitrarily assuming that $P_{ig} > P_{jh}$, we define a Guttman error as a $(0, 1)$ score pattern with $Y_{ig} = 0$ on the easiest item step and $Y_{jh} = 1$ on the most difficult item step. A person with total score $X_+ = X_i + X_j = x$ on the item pair, where $x = 0, 1, \ldots, 2m$,

Table 7.2   Univariate Item Step Popularities and Item Means for 828 Respondents and Three Items, Scored 0 (Never), 1 (Rarely), 2 (Often), and 3 (Always or Almost Always).

Item Number	Contents	Mean	Item Step Proportions		
			≥ 1	≥ 2	≥ 3
1	Keep windows closed	1.86	.885	.600	.379
2	No laundry outside	1.38	.662	.435	.280
4	No blankets outside	1.33	.604	.426	.301

has no Guttman errors if he or she has passed the x easiest item steps and failed the remaining $2m - x$ more difficult ones. If this is not the case, one may count and weigh the Guttman errors. This counting and weighing is at the basis of calculating the H coefficients for the polytomous item case, but we consider the detailed account of it beyond the scope of this book and refer the interested reader to Molenaar (1991) and Molenaar and Sijtsma (2000, pp. 20–21) for details. Here we present an illustration, however, for which we use the industrial malodor data based on Table 7.2.

A Numerical Example. From the first two lines in the body of Table 7.2, it can be deduced that the ordering of the six item steps from easy to difficult is: $X_1 \geq 1, X_2 \geq 1, X_1 \geq 2, X_2 \geq 2, X_1 \geq 3, X_2 \geq 3$. Table 7.3 gives the cross-tabulation of Item 1 and Item 2. There are 63 persons with item scores (0, 0), who thus have total score $X_+ = X_1 + X_2 = 0$ on the item pair. It follows from Table 7.2 that the step $X_1 \geq 1$, passed by a fraction $733/828 = .885$ of the respondents, is the easiest item step. So the 79 persons in the (1, 0) cell of Table 7.3 with $X_+ = 1$ have no Guttman error, whereas the 19 persons in the (0, 1) cell have one Guttman error: They passed $X_2 \geq 1$, with a popularity of $548/828 = .662$ (also, see Table 7.2), but failed the easier step $X_1 \geq 1$.

Proceeding in this way, we find seven cells, marked by dashes, in which there are no Guttman errors. Cell (0, 3), however, contains 10 persons who each have six weighted Guttman errors: Based on the six item steps from easy to difficult, $X_1 \geq 1, X_2 \geq 1, X_1 \geq 2, X_2 \geq 2, X_1 \geq 3, X_2 \geq 3$, the persons in cell (0, 3) score (0, 1, 0, 1, 0, 1) on these steps, and this pattern has six occasions on which a 0 precedes a 1. In the weighted count (not further explained) of observed errors across the whole table, we find a total of

Table 7.3   Cross-Tabulation of Scores on Items 1 and 2 and Calculation of Pairwise H; Each Cell Gives Observed Frequency, Expected Frequency, and Weight of Guttman Errors; Also, Marginal Distributions and Cumulative Marginal Distributions Are Given, as Well as Totals Leading to H Coefficient.

		X_2					
X_1		0	1	2	3	*Marginal*	*Cumulative*
0	Observed frequency	63	19	3	10	95	828
	Expected frequency	32.1	21.6	14.7	26.6		$X_1 \geq 0$
	Weight of errors	–	1	3	6		
1	Observed frequency	79	105	28	24	236	733
	Expected frequency	79.8	53.6	36.5	66.1		$X_1 \geq 1$
	Weight of errors	–	–	1	3		
2	Observed frequency	46	38	64	35	183	497
	Expected frequency	61.9	41.6	28.3	51.3		$X_1 \geq 2$
	Weight of errors	1	–	–	1		
3	Observed frequency	92	26	33	163	314	314
	Expected frequency	106.2	71.3	48.5	88.0		$X_1 \geq 3$
	Weight of errors	3	1	–	–		
	Marginal	280	188	128	232		
	Cumulative	828	548	360	232		

Sum Observed	Sum Expected	St. Dev.	H	Z
571	963.20	36.7133	0.4072	10.6829

$$F = 19 \times 1 + 3 \times 3 + 10 \times 6 + 28 \times 1 + 24 \times 3$$
$$+ 46 \times 1 + 35 \times 1 + 92 \times 3 + 26 \times 1 = 571. \qquad (7.12)$$

Similar to the dichotomous case, this observed count must be compared to the expected count E under global independence given the observed marginals. The expected counts can be obtained from the marginal distributions of the table, just as when a chi-square test is performed. For example, for cell (0, 1), the expected count is found to be $95 \times 188/828 = 21.6$. In general, the numbers in the middle line of each cell give the expected count for that cell. Thus, we obtain that $E = 963.20$ and $H_{1,2} = 1 - 571/963.2 = 0.4072$. Under the null hypothesis of global independence, and given the observed marginals, F would have expectation E, standard deviation 36.7 (not derived here; it follows from the multinomial distribution of the 16 entries in the 4×4 table), and a Z

score of $|571 - 963.2|/36.7 = 10.68$. Thus, the value $H = 0.41$ found in the sample of 828 respondents permits us to reject beyond any doubt the null hypothesis that the pairwise H would be 0 in the population.

As in Equation 4.4 for the dichotomous case, for polytomous items, the pairwise H is equal to the observed covariance divided by the maximum possible covariance given the marginals, and to the similar ratio of the correlations. Thus, an alternative way to calculate the pairwise H is by means of covariances or correlations.

Equation 4.5 for the scalability of an item with respect to the other $k - 1$ items, denoted H_i, and Equation 4.7 for the scalability of the total item set, denoted H, hold also for the polytomous item case. From a table similar to Table 7.3 but not given here, one finds $F = 629$ and $E = 1,023.06$ for item pair $(1, 4)$. Because our scale has only three items, this implies that the scalability of Item 1 with respect to the other two items is given by

$$H_1 = 1 - \frac{F_{1,2} + F_{1,4}}{E_{1,2} + E_{1,4}} = 1 - \frac{571 + 629}{963.20 + 1023.06} = 0.40. \quad (7.13)$$

Moreover, the item pair $(2, 4)$ turns out to have $F = 236$ and $E = 1,242.75$, and for the three-item scale, this implies

$$H = 1 - \frac{F_{1,2} + F_{1,4} + F_{2,4}}{E_{1,2} + E_{1,4} + E_{2,4}}$$

$$= 1 - \frac{571 + 629 + 236}{963.20 + 1023.06 + 1242.75} = 0.56. \quad (7.14)$$

We remind the reader that Chapter 4 ended with some useful properties of the scalability coefficients and a discussion of their interpretation. Chapter 5 introduced an automated item selection procedure. All this material applies without change to the polytomous case.

Goodness of Fit: Monotonicity Assumption

In the dichotomous case, the item-rest regression (see Chapter 3) was used as an indirect way to verify the monotonicity of the IRF. In an exactly analogous way, the restscore on the remaining $k - 1$ polytomous items,

$$R_{(i)} = \sum_{j=1; j \neq i}^{k} X_j, \quad (7.15)$$

will be used (see also Equation 3.1). In practical data analysis, a suitable grouping of adjacent restscore values is done so that enough power is obtained to verify monotonicity for each of the m ISRFs of a particular item. When the number of violations is low and they are of a modest size, there is no reason to doubt the monotonicity of the ISRFs. If violations are numerous and/or large, the item currently examined may have to be removed from the scale, because it might disturb the ordering of the respondents by means of X_+. It is our experience that items for which the monotonicity assumption is violated often have low scalability coefficients, H_i, which then fall below the default value of the lowerbound $c = 0.3$. In Chapter 8, we consider in detail the itemstep-rest regressions of Item 1 (rest score based on Items 2 and 4 for the three-item scale from the industrial malodor data) and try to explain why in this unusual case several serious violations of monotonicity still go together with a value as high as $H_1 = 0.40$.

As before, when the monotonicity assumption is violated to such an extent that items should be removed, it is recommended to do this item by item, because removal of an item often leads to higher item scalability coefficients and to fewer item-restscore violations for the remaining items. Also, item removal should not be automatic, because item content should also play a role. For example, when the contents of an item do not give a clue as to why the monotonicity of the item may be violated, and the number of available items is small, one might decide to keep the item in the test after all, in order to have better coverage of the latent trait. Finally, especially for polytomous items with their fine-grained total scoring from 0 to $m \times k$, there tend to be many possible values of the restscore, several of which may have very low observed frequencies. Because combining them in different ways may easily either mask or exaggerate violations of monotonicity, one is advised to consider several groupings before deciding whether monotonicity is systematically violated. This can be done conveniently by means of the MSP software (Molenaar & Sijtsma, 2000).

The violation counts, which are implemented in MSP, are used in a descriptive manner. For large data sets, one should not be disturbed by the occurrence of many violations, a phenomenon that is also known from fit studies for parametric IRT models. The probability distributions of the number and size of violations under the null hypothesis that the MHM holds in the population are very complicated, and as a result, we have no formal overall significance checks.

Comparison With Parametric Polytomous IRT Models

The MHM can be seen as a nonparametric version of Samejima's (1969, 1972) homogeneous case of the graded response model. That model defines the ISRF as a logistic function (Chapter 2) with a slope (α_i) and a location (λ_{ih}) parameter. For a fixed item, the slope parameter is the same for each ISRF,

$$P(X_i \geq h|\theta) = \frac{\exp[\alpha_i(\theta - \lambda_{ih})]}{1 + \exp[\alpha_i(\theta - \lambda_{ih})]}, \tag{7.16}$$

for $h = 0, \ldots, m$; and $P(X_i \geq 0|\theta) = 1$. The homogeneous case of the graded response model is a well-known parametric IRT model from the class of *cumulative probability models*, with nondecreasing ISRFs generally defined as

$$f_{ih}(\theta) = P(X_i \geq h|\theta). \tag{7.17}$$

Unidimensional measurement, local independence, and $P(X_i \geq x|\theta)$ (Equation 7.17) nondecreasing in θ define the MHM exactly. Hemker, Sijtsma, Molenaar, and Junker (1996) have called the MHM the nonparametric graded response model to emphasize the close relationship between Samejima's graded response model and the MHM.

Van Engelenburg (1997) argued that the graded response model (and, in fact, all cumulative probability models) is best suited for modeling item scores that result from a global assessment task, for example, the rating of a response on a Likert-type item measuring an attitude or a personality trait. Here, the respondent forms a general impression of his or her position on the scale relative to the item. For example, the respondent is asked to determine on a Likert scale the degree to which he or she considers a situation to be in agreement with his or her own typical behavior, and the psychologist then may interpret the response given as indicating a particular degree of extroversion.

It has been shown (Hemker, Van der Ark, & Sijtsma, 2001) that of *all* the well-known polytomous IRT models (i.e., not only the cumulative probability models, but also others), the MHM is the most general model. All other models are special cases. This means that if any of these models, for example, the parametric graded response model (Equation 7.16), fits the data, then the MHM also fits the data. The reverse situation is more important: If the MHM fits the data, but the parametric graded response model (or another model) does not, then we may still use total score X_+

for ordering persons on θ. This result gives additional justification for the use of the MHM.

Goodness of Fit: Assumption of Nonintersecting ISRFs

The DMM for polytomous items is a special case of the MHM for polytomous items. The DMM assumes unidimensional measurement, local independence, monotone nondecreasing ISRFs, and, *in addition*, that the ISRFs of different items do not intersect. That is, if we know for one value of the latent trait, say, θ_0, that for two items, i and j, $P_{ig}(\theta_0) < P_{jh}(\theta_0)$, then we know for any θ that

$$P_{ig}(\theta) \le P_{jh}(\theta), \tag{7.18}$$

and similar orderings hold for all $m \times k$ ISRFs from the test (these orderings are structural for ISRFs from the *same* item). Figure 7.2 shows two items with four ISRFs each, of which none intersect.

Because the DMM is a special case of the MHM, it also allows us to use X_+ as a proxy for ordering respondents on θ. Also, it is easily seen that for each θ, all ISRFs have the same ordering, with the exception of possible ties. This might be termed an invariant item step ordering.

In Chapter 6, several fit procedures for the DMM were described for dichotomous items. Each of these methods has been generalized to polytomous items, thus making possible the investigation of the intersection of ISRFs of different items. Also, the classical reliability coefficient was estimated via interpolation in the $\mathbf{P}(++)$ matrix (see Equation 6.11a) containing the fractions of respondents answering positively to two items. Reliability estimation also has been generalized to polytomous items, using a version of the $\mathbf{P}(++)$ matrix for item steps (not further discussed here).

A complete polytomous DMM item analysis can be done using the generalized methods described in this chapter, which have been incorporated in the MSP software, with the exception of the H^T coefficient. We will not go farther into the polytomous DMM item analysis here because, for one reason, in practice such an item analysis may be rather complex due to its voluminous output. More important, however, is the observation that the DMM focuses on the nonintersection of the ISRFs, whereas practical researchers tend to focus on the whole item as the unit of interest. That is, they may be less interested in the ordering of the ISRFs than in the ordering of the items as a whole. This has led us to

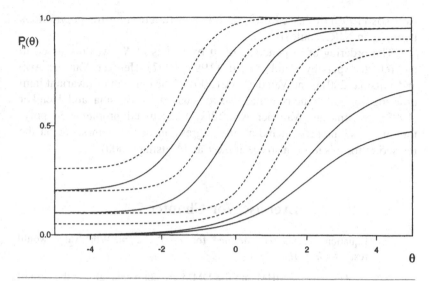

Figure 7.2 Nonintersecting Item Step Response Functions for Two Items With Five Ordered Answer Categories Each.

concentrate our research on polytomous IRT with the item as the unit of interest rather than the item step and the ISRF. Chapter 8 gives a real data example that investigates the ordering of items in relevant subgroups, and that has the item as the unit of interest. The theory supporting polytomous IRT with the item as the central unit is currently under investigation.

Additional Reading

The polytomous MHM was discussed by Sijtsma, Debets, and Molenaar (1990), Hemker, Sijtsma, and Molenaar (1995), and Molenaar (1982c, 1997). Mathematical background of a similar model, in particular focusing on person measurement, was provided by Junker (1991). Molenaar (1991) discussed the theory of the H coefficient for polytomous items, and Molenaar and Sijtsma (1988) generalized reliability estimation for X_+ to polytomous items. Vermunt (2001) discussed a general approach based on latent class analysis to testing a wide range of polytomous IRT models, including the models discussed here. Hemker et al. (1995) discussed results of applying the automated item selection algorithm to sets of polytomous items. Chang, Mazzeo, and

Roussos (1996) discussed differential item functioning for polytomous items in an NIRT context.

The ordering of respondents on θ by means of X_+ was discussed in theoretical papers by Hemker et al. (1996, 1997), Hemker, Van der Ark, and Sijtsma (2001), and Van der Ark (2001). The concept of invariant item ordering for polytomous items was discussed by Sijtsma and Hemker (1998). Sijtsma and Van der Ark (2001) discussed problems in polytomous NIRT models, including the regression of the item score on the restscore; for this topic, also see Junker and Sijtsma (2000).

Exercises for Chapter 7

7.1. Equation 7.2 has been defined for $h = 0, \ldots, m$. What value would result for $h = 0$?

7.2. Explain why the MHM and the DMM are not meaningful in the case of items with $m > 2$ nominal answer categories.

7.3. Which order property can be proven for dichotomous items but may be false for polytomous items?

7.4. Can two items with a negative covariance belong to the same scale that satisfies the MHM?

7.5. Indicate whether each of the following aspects is the same or different when one passes from dichotomous to polytomous items; if different, say how.
   a. Stochastic order of latent trait value by total score   same/different
   b. MHM implies nonnegative covariances   same/different
   c. Items ordered by item means (under DMM)   same/different
   d. Use of $\mathbf{P}(++)$ and $\mathbf{P}(--)$ matrices for assessing DMM same/different
   e. Assumption of local independence   same/different
   f. Calculation of H for an item pair   same/different
   g. Automatic item selection based on H values   same/different

7.6. Suppose that you join two adjacent categories in the data measuring coping with industrial malodor. Do you expect the item H values before and after joining to be exactly the same, slightly different, or quite different? Explain why.

7.7. The Disability Assessment Schedule (DAS) is a scale on which the dysfunctioning of psychiatric patients is rated. Based on a semistructured interview with someone who knows the patient well, a psychiatrist or nurse rates the patient's performance on seven activity areas by 0 (no problems), 1 (light dysfunctioning), or 2 to 5 (more serious dysfunctioning; in our data set, these categories were combined and scored as 2 because of low frequencies). For the Items DAS2 (underactivity) and DAS6 (sexual role), it was found that among 236 patients, 63 were in the (0, 0) cell, 35 in (0, 1), 29 in (0, 2), 13 in (1, 0), 12 in (1, 1), 23 in (1, 2), 7 in (2, 0), 6 in (2, 1) and 48 in (2, 2).

a. Write out the 3×3 cross-tabulation with exact and cumulative marginal entries.

b. Order the four item steps from easy to difficult.

c. Mark by dashes the path from (0, 0) to (2, 2) of cells without Guttman errors.

d. How many weighted Guttman errors are there in each cell that is not on the path?

e. Calculate the observed weighted error count F and its expectation E under the null hypothesis of global independence.

f. Find the table with the same marginals for which the covariance is maximal (Hint: This is the table without Guttman errors).

g. Calculate the pairwise H for DAS2 and DAS6, both as $1 - F/E$ and as Cov/Cov_{max}.

h. Join categories 1 and 2 into a new category 1 and repeat Steps e, f, and g.

Answers to Exercises for Chapter 7

7.1. $P_{i0}(\theta) = P(X_i \geq 0|\theta) = 1$, for each i and each θ.

7.2. Both models order the persons by their sum score and the items by their item means; neither is meaningful if the category scores $0, \ldots, m$ are nominal rather than ordinal.

7.3. Stochastic ordering on the latent trait value θ by the sum score X_+.

7.4. No, the MHM implies nonnegative covariances. Because the model assumptions refer to the population and not to the sample, a sample covariance might be negative due to sampling error even though the population value is still positive, but it would be hazardous to count on that explanation.

7.5. a. different (stochastic order proven for dichotomous items, counter-examples exist for polytomous items)

b, d, e, and g, same

c. different (for dichotomous items, we have an IIO, and for polyto-mous items an invariant ordering of item steps, but not of items)

f. different (more than one error cell, weighted count)

7.6. After joining, both F and E are usually lower because errors between the categories now joined become invisible. It is therefore hard to say exactly what pairwise H values will do. Except for extreme frequencies, it is plausible that the effect on the item H values will be minor, but it will be rare that the numerical values remain exactly the same.

7.7. a, c, and d, see Table 7.5
  b. DAS6 $\geq$ 1, DAS2 $\geq$ 1, DAS6 $\geq$ 2, DAS2 $\geq$ 2.
  e. $F = 29 + 13 + 7 \times 3 + 6 = 69$ and $E = 53.8 + 16.9 + (21.5 \times 3) + 13.7 = 148.9$.
  f. Put zeros in the four error cells. Given the fixed marginals of the table, the other five frequencies follow by implication: see Table 7.4
  g. $H = 1 - 69/148.9 = 0.54$, Cov = .6085, $\text{Cov}_{max} = 1.1349$, $r = .4543$, $r_{max} = .8473$.
  h. $F = 20$, $E = 38.33$, $H = 0.48$, Cov = .0777, $\text{Cov}_{max} = .1624$, $r = .3264$, $r_{max} = .6823$.

Table 7.4

83	44	0	127
0	9	39	48
0	0	61	61
83	53	100	236

Table 7.5

DAS2		DAS6 0	1	2	Total	Cumulative
0	obs	63	35	29	127	236
	exp	44.7	28.5	53.8		
	wgt	–	–	1		
1	obs	13	12	23	48	109
	exp	16.9	10.8	20.3		
	wgt	1	–	–		
2	obs	7	6	48	61	61
	exp	21.5	13.7	25.8		
	wgt	3	1	–		
Total		83	53	100	236	
Cumulative		236	153	100		

8

Item Analysis Using Nonparametric IRT for Polytomous Items

In Chapter 7, we explained that the MHM and the DMM for dichotomous items can be generalized to polytomous items. The goal of this chapter is to illustrate how these models can be used to analyze polytomous item scores, and how the results should be interpreted. The following topics are discussed. First, we test as one scale the data set from the questionnaire of 17 four-category items on coping with industrial malodor in the vicinity of one's home (Cavalini, 1992; shorthand item contents in Table 8.1). The results demonstrate that this questionnaire is not unidimensional. Second, based on the application of the automated item selection procedure described in Chapter 5, we pick out one subscale and look in detail at the itemstep-rest regression for polytomous items.

The final section deals briefly with the analysis of five-category item scores from a questionnaire asking Dutch labor union members about their willingness to participate in each of six kinds of protest action, such as attending a union meeting during working hours, marching in a demonstration for improved working conditions, or participating in a strike. The sample consisted of members from several divisions of the union, such as the publishing industry, housing construction, civil service, and transportation. This gave us the opportunity to compare the scalability of the data and the ordering of the means of action in different groups of respondents.

Table 8.1   Mean, Standard Deviation (SD), Scalability (H_i), and Number of Significant Violations of Itemstep-Rest Regression (#SiVi) for Two Minimum Group Sizes [#SiVi(. . .)] of 17 Items Measuring Coping With Industrial Malodors.

Number	Item Text	Mean	SD	H_i	#SiVi (20)	#SiVi (100)
1	Keep windows closed	1.86	1.05	.22	2	–
2	No laundry outside	1.38	1.21	.18	7	1
3	Search source of malodor	1.85	1.05	.21	1	–
4	No blankets outside	1.33	1.27	.19	2	–
5	Try to find solutions	.82	.95	.20	–	–
6	Go elsewhere for fresh air	.54	.71	.20	–	–
7	Call environmental agency	.26	.63	.18	–	–
8	Think of something else	.76	.85	.17	1	1
9	File complaint at producer	.35	.70	.14	1	–
10	Acquiesce in odor annoyance	1.56	1.11	–.06	113	17
11	Do something to get rid of it	.86	.93	.19	1	–
12	Say: It might have been worse	1.07	.92	–.07	117	20
13	Experience unrest	.65	.79	.20	–	–
14	Talk to friends and family	.98	.84	.18	5	–
15	Seek diversion	.78	.81	.22	–	–
16	Avoid breathing through the nose	.62	.90	.15	3	–
17	Try to adapt to situation	1.95	.96	.05	11	6

Analysis of Coping Behavior Questionnaire as One Scale

Each of the 17 four-category items (Table 8.1) is a possible reaction to the question, "What do you do or think when you smell industrial malodor?" Each item was scored 0, 1, 2, or 3, with a higher score indicating a more frequent use of this strategy and thus possibly a stronger inclination to exhibit a particular coping behavior. Table 8.1 gives the item means and standard deviations based on data from 828 respondents living in the vicinity of a factory creating bad-smelling air. Item 7 ("call environmental agency") and Item 9 ("file complaint with producer") both express a highly active and perhaps even daring attitude, which may explain their low group means, indicating that relatively few respondents say they will take this course of action. None of the items' standard deviations seem to be remarkably low or high.

For the 17 items, 136 item pair scalability values H_{ij} were computed, 30 of which were negative, which must not occur under the MHM (Chapter 7). Ten of the 16 H_{ij}s of Item 10 were negative; 13 H_{ij}s of Item 12 were negative; and 7 H_{ij}s of Item 17 were negative. None of the other items had more than three negative H_{ij}s. Based on these descriptive statistics alone, attention is quickly drawn to Items 10 ("acquiesce in odor annoyance"), 12 ("say: It might have been worse"), and 17 ("try to adapt to situation").

Individual item H_is with respect to the other 16 items were calculated for each of the 17 items. Two H_is were negative (Table 8.1). The H_i values of each of the other 15 items (Table 8.1) were all significantly larger than 0 at the 0.003 significance level (with $\alpha = 0.05$; using the Bonferroni rule we have $0.05/15 \approx 0.003$, and a one-sided $Z_{crit} \approx 2.72$; the smallest absolute Z was $Z_{17} = 4.57$). However, 14 positive H_i values were between 0.14 and 0.22, and one H_i was equal to 0.05. All H_i values were lower than the commonly accepted lowerbound 0.3 (Chapters 4 and 5). Because for the total H we know that $H \leq \max(H_i)$ (Chapter 4, Theorem 4.2), we know that for these data, $H \leq H_1 = H_{15} = 0.22$ (Table 8.1); in fact, for all 17 items, $H = 0.15$.

Although the low H_is are reason enough to proceed directly to a search using the automated item selection procedure, we first take a brief look at results for the itemstep-rest regressions. An analysis of the 17×3 itemstep-rest regressions (a detailed example is presented in the next section) with a minimum restscore group size of 20 respondents, thus estimating many points of the itemstep-rest regressions but with little accuracy, yielded large numbers of significant violations for Item 10 (also note that $H_{10} = -0.06$) and Item 12 (also, $H_{12} = -0.07$), but zero to moderate numbers for the other items (Table 8.1). Choosing a minimum restscore group size of 100, thus estimating few points of the itemstep-rest regression but with high accuracy, yielded almost no significant violations for 14 items, but Items 10, 12, and 17 still stood out with 17, 20, and 6 significant violations, respectively. Items 10 and 12 express resignation to the inconvenient situation, which also could be said of Item 17 ($H_{17} = 0.05$). Based on scalability coefficients and itemstep-rest regressions, it seems that the other 14 items have more in common with each other than with Items 10, 12, and 17. Given their low H_i values, however, it is likely that these 14 items do not constitute just one scale. A more detailed item analysis will reveal this.

A Closer Look at Itemstep-Rest Regression
for Polytomous Items

In Chapter 5, we used the automated item selection procedure for analyzing a dichotomized version of the polytomous coping scores. We used coefficient H with several lowerbound c values ranging from 0.00 to 0.60 in 0.05 unit increments. For this chapter, we also analyzed the polytomous item scores, and the results of the automated item selection procedure were highly similar to those obtained for dichotomous items. Because of this similarity, we refrain from a tedious repetition of almost the same discussion, and consider only some interesting results for two scales that resulted from the item selection.

For lowerbound $c = 0.0$ and $c = 0.3$, Items 10, 12, and 17, which were interpreted to measure resignation, came out as a separate scale with $H = 0.35$, and item coefficients $H_{10} = 0.35$, $H_{12} = 0.35$, and $H_{17} = 0.36$. In total, there were 18 violations of nondecreasingness of the 3×3 itemstep-rest regressions, two of which were significant. The size of these significant deviations was 0.09 (Item 10) and 0.13 (Item 17). Figure 8.1 shows the three itemstep-rest regressions for Item 10 and serves to give the reader a first impression of how these regressions look when the

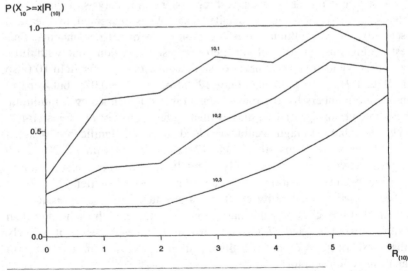

Figure 8.1   Estimated Itemstep-Rest Regressions of Item 10.

MHM fits the data by approximation. That is, the regressions are *monotone* except for some small and unimportant deviations. Notice that the regressions actually are discrete points that we have connected by straight lines and that they estimate continuous ISRFs of the kind shown in Figure 7.1. We discuss itemstep-rest regressions in more detail below when we consider a subscale for which several regressions are clearly *nonmonotone*. Given the rules of thumb for the interpretation of H, the Resignation scale is a weak scale. This qualification is further corroborated by the moderately sized significant violations of nondecreasingness of the itemstep-rest regressions.

All remaining 14 items had positive covariances. Because the definition of a scale requires that all covariances are positive and all item scalability coefficients H_i at least equal to c, for $c = 0.0$, these 14 items thus were all selected into the same scale. For $c = 0.3$, this item set was split into three subsets. We now take a closer look at one of them, the scale consisting of Items 1, 2, and 4. In particular, we consider in detail the itemstep-rest regressions as proxies for the ISRFs.

For Items 1, 2, and 4 as one scale, $H = 0.56$, with item scalability coefficients $H_1 = 0.40$, $H_2 = 0.63$, and $H_4 = 0.62$. This is a strong scale, although the H_i value of Item 1 ("keep windows closed") is lower than those of Item 2 ("no laundry outside") and Item 4 ("no blankets outside"). As explained in Chapter 7, an item with four ordered answer categories is characterized by three meaningful ISRFs, defined as $P_{ih}(\theta) = P(X_i \geq h|\theta)$ (Equation 7.2), with $h = 1, 2, 3$. The 3×3 itemstep-rest regressions of Items 1, 2, and 4 are displayed in Figure 8.2. These itemstep-rest regressions are considered to be estimates of the ISRFs.

Figure 8.2 provides strong indications that in general the monotonicity assumption is not supported by the data. Table 8.2 provides the numerical results for the three itemstep-rest regressions of Item 1. With three four-category items in our scale, the restscore $R_{(i)}$ is based on two items and runs from 0 to 6. The item means (Table 8.2) for restscore groups do not always increase with $R_{(1)}$ as one would expect. It may be noted that in a fixed restscore group, the item mean is the sum of the three proportions $P(X_i \geq h|\theta)$ across $h = 1, 2, 3$, which is easily checked in Table 8.2; for example, for $R_{(1)} = 0$, we have $1.61 = 0.77 + 0.49 + 0.35$. It follows that we will encounter decreases in the itemstep-rest regressions displayed next to the column with the item means.

Because seven restscore groups yield $7 \times 6/2 = 21$ opportunities to assess pairs, for each itemstep-rest regression 21 pairs of proportions can

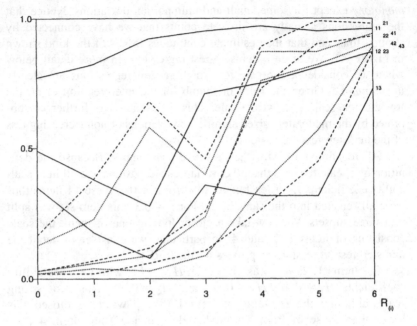

Figure 8.2   Estimated Itemstep-Rest Regression of Item 1 (Solid Lines), Item 2 (Dashed Lines), and Item 4 (Dotted Lines). Notation "11" to the right of the figure is shorthand for "$P(X_1 \geq 1|R_{(1)})$"; "42" stands for "$P(X_4 \geq 2|R_{(4)})$"; and so on.

provide evidence either for or against monotonicity. It may be noted that in general, pairs that have 0.00 in the first position or 1.00 in the second position cannot provide evidence against monotonicity and, therefore, cannot be considered sources of information on violations of monotonicity. The other pairs are called active pairs, because they do contain such information. For our data, the lower half of Table 8.2 shows that all 21 pairs are active pairs ($\#ap = 21$). The number of violations ($\#vi$) and the mean number of violations ($\#vi/\#ap$) are largest for $h = 3$. The more serious violations are found for $h = 2$ and $h = 3$, which can be seen from the sum of all violations for each value of h, the mean violation ($sum/\#ap$), and the maximum violation ($maxvi$) found (for $h = 3$, this is 0.27, occurring between groups 0 [0.35] and 2 [0.08]; see heading $group$). Moreover, based on a standard normal test (maximum Z values under heading $zmax$),

Table 8.2 Three Itemstep-Rest Regressions of Item 1 ($P[X_1 \geq h|R_{(1)}]$; Upper Half of Table); and Descriptive Results and Test Results for Violations of Monotonicity (Lower Half).

Restscore		Item Score Frequencies				Item Mean	$P[X_1 \geq h|R_{(1)}]$		
$R_{(1)}$	n	0	1	2	3		≥ 1	≥ 2	≥ 3
0	239	55	66	35	83	1.61	0.77	0.49	0.35
1	85	15	38	17	15	1.38	0.82	0.38	0.18
2	118	10	75	24	9	1.27	0.92	0.28	0.08
3	58	3	14	20	21	2.02	0.95	0.71	0.36
4	92	3	20	40	29	2.03	0.97	0.75	0.32
5	37	2	5	18	12	2.08	0.95	0.81	0.32
6	199	7	18	29	145	2.57	0.96	0.87	0.73

$P[X_1 \geq h|R_{(1)}]$	#ap	#vi	#vi/#ap	maxvi	sum	sum/#ap	zmax	group	#zsig
≥ 1	21	3	0.14	0.02	0.03	0.00	−0.33	4 6	0
≥ 2	21	3	0.14	0.21	0.43	0.02	3.79	0 2	2
≥ 3	21	7	0.33	0.27	0.68	0.03	5.88	0 2	3
Item Total	63	13	0.21	0.27	1.14	0.02	5.88		5
ItTo,$vi \geq 0.03$	63	9	0.14	0.27	1.09	0.02	5.88		5
ItTo,$group \geq 100$	18	2	0.11	0.23	0.40	0.02	5.64		2

for $h = 2$, there are two significant violations (#*zsig*), and for $h = 3$, there are three significant violations. It can be concluded that there is strong evidence that these two ISRFs do not satisfy monotonicity. Given this result, one may ask why coefficient H and the H_is have high values. The answer is that for X_+ based on the three Items 1, 2, and 4, the frequency distribution of X_+ has a large spread (Table 7.1).

The MSP program has several options for manipulating the granularity of the monotonicity investigation. For example, the next-to-last line of Table 8.2 shows total results for a minimum group size of 20, as before, but now violations are counted only when they are at least 0.03 (line ItTo, $vi \geq$ 0.03; the value of 0.03 is the default of MSP). Because the itemstep-rest regressions still show quite a number of larger violations, this manipulation can be seen to have little effect on the results. The last line of Table 8.2 shows total results in which, in addition, the minimum restscore group size is 100 instead of 20. Thus, small sample violations (up to 0.03) are ignored and only a few points of the itemstep-rest regressions are (accurately) estimated. Forcing restscore groups to be large has the effect of producing a total of only four groups (reflected by #ap), and as a result, fewer violations are found (see #vi and sum), two of which are significant.

Given the interpretation that H gives information about the degree to which respondents can be ordered by means of X_+, and given the rather strong violations of monotonicity as evidenced by the itemstep-rest regressions that seem to contradict this interpretation, what should our conclusion be? The answer is that the scale allows for the ordering of respondents who are far apart (low X_+ values vs. high X_+ values), but that little can be said about the ordering of respondents who are closer together on the scale. Not only does random measurement error disguise the true ordering of these respondents, but the nonmonotonicity of the ISRFs introduces added systematic distortions in the ordering. Thus, measurement is further impaired, and the effect is stronger the closer together respondents are located on the scale. This conclusion could be drawn only by combining information about the distribution of X_+, the itemstep-rest regressions, and the H coefficients.

Investigating the Nonintersection of ISRFs

In Chapter 7, it was pointed out that the DMM for polytomous items differs from the MHM by the additional assumption of nonintersecting ISRFs. Clearly, a DMM analysis of polytomous data yields more voluminous output than that of dichotomous data. For example, for $(m + 1)$-category items, the $\mathbf{P}(++)$ and $\mathbf{P}(--)$ matrices have $k \times m$ rows and columns, whereas for dichotomous items, this number is only k. Similarly, the restscore method (Chapter 6) is more laborious for polytomous items because many more itemstep-rest regressions have to be compared between items. Because of the complexity of a polytomous DMM analysis, we refrain from a detailed example and instead refer the reader to a few publications where interesting data analyses using the polytomous DMM can be found (see "Additional Reading").

Scalability and Item Ordering in Different Groups

The next data example illustrates how one can investigate the scalability of an item set and the ordering of the items for relevant subgroups of respondents. Differences in scalability between subgroups may be an indication that the test measures different latent traits. Different item orderings between subgroups can be indicative of DIF, as was explained in detail in Chapter 6. Also, different item orderings may help to explain lack of person fit in a particular subgroup. Finally, as the next example shows, knowledge of the item ordering within several interesting subgroups may contribute to the development of theory about certain behaviors.

Van der Veen (1992) studied the willingness of Dutch labor union members to participate, in principle, in different means of action directed toward achieving a goal that cannot be reached through regular negotiations between, for example, company management and union representatives. Goals may be better wages, more jobs, better job opportunities, shorter hours, and safer or healthier working conditions. One part of Van der Veen's study dealt with the analysis of ordered five-category questionnaire data concerning attitudes about means of action obtained from a sample of members from different divisions of the union. Examples are the publishing industry, housing construction, civil service, and transportation. She was interested in, among other things, the ordering of actions in different union divisions.

Here we reanalyze only a part of Van der Veen's questionnaire data, again using the MSP software. The goal is to compare six actions across different union divisions according to (a) total and individual action scalability and (b) ordering by mean score. The sample consisted of 496 union members, divided across nine union divisions (a short description of the divisions and the frequencies for each division can be found in Table 8.3). Each sampled union member rated on a five-point scale his or her general attitude toward each of six actions: (a) *Strike* (a complete cessation of work; no productivity); (b) working according to official *Regulations* (meaning no flexibility or adaptation to the demands of the client, resulting in lower productivity); (c) *Interruption* of work to *Demonstrate* the potential power of a union to mobilize its members; (d) *Interruption* of work to *Communicate*, for example, by means of a meeting, information to workers or management officials; (e) *Protest Meeting* outside of working hours to exchange information; and (f) *Demonstration* outside of working hours to emphasize the power of the union and its members.

Scaling of Action Means in the Whole Group and in Divisions

In the *whole group*, given 15 covariances among six items, only the covariance between Strike and Protest Meeting was negative. These

Table 8.3 Means Based on Six Items in Total Group and in Nine Union Divisions (Group Sizes at Bottom of Table), Scalability Coefficients H_i for Individual Items and for Total Scale.

Item Contents	Total Group H_i	1	2	3	4	5	6	7	8	9
					Union Divisions					
Strike	1.38	.43	1.5	1.3	1.3	1.1	1.6	1.5	1.4	1.5	1.2
Interrupt for Communication	2.16	.48	2.1	2.4	1.9	1.8	2.5	2.1	2.1	2.1	2.2
Demonstration	2.21	.36	2.4	2.3	2.2	2.0	2.0	2.1	2.2	2.5	1.8
Interrupt for Demonstration	2.27	.55	2.6	2.4	1.9	2.0	2.3	2.3	2.2	2.2	2.3
Regulations	2.28	.51	2.6	2.4	2.0	1.7	2.2	2.2	2.6	2.4	2.1
Protest Meeting	2.65	*	2.6	2.7	2.6	2.7	2.6	2.7	2.6	2.8	2.5
Group Size	496		58	100	52	41	45	52	51	52	45
Total Scale H (5 items)		.47	.46	.49	.49	.45	.54	.52	.51	.55	.43

*Rejected from item selection due to negative covariance with Strike.

Union Divisions (no official names; only indications are given): (1) Police; (2) Civil Service; (3) Public Services; (4) Publishing Industry; (5) Housing Construction; (6) Industry and Nutrition; (7) Teachers (Catholic); (8) Teachers (Protestant); (9) Transportation.

items thus cannot be part of the same MHM scale. Strike had the lowest mean and Protest Meeting the highest mean of all six actions (Table 8.3). Thus, Strike was rated least favorably and Protest Meeting most favorably. Application of the automated item selection procedure (Chapter 5) with lowerbound $c = 0.3$ yielded a scale consisting of five items ($H = 0.47$) with individual scalability values between 0.36 and 0.55 (Table 8.3). Protest Meeting was rejected due to its negative covariance with Strike. Its scalability value with respect to the five selected items was 0.09.

Application of the automated item selection procedure also led to the rejection of Protest Meeting in eight out of nine *divisions*. The only exception occurred for the Publishing Industry division, where Protest Meeting and Demonstration were selected first because they had the highest H_{ij}, and Strike was rejected because of its negative covariance with Protest Meeting. The end result for this division, however, was that Protest Meeting had an H_i of only 0.2 (rounded) with respect to the other four selected items, whereas the rejected Strike had an H_i of 0.4 (rounded) with respect to the five selected items (typical findings like this are explained in Chapter 5). Correcting the item selection by rejecting Protest Meeting and including Strike thus seems reasonable. In three divisions, only four items were selected into the first scale, and two other items (Protest Meeting and Demonstration) were selected into a second scale. The total H of the first scale across the nine divisions ranged from 0.43 to 0.55 (Table 8.3). We refrain from a more detailed interpretation of results because division sample sizes were small (Table 8.3).

Item Ordering in Divisions of the Labor Union

The ordering of the ISRFs was studied in the whole group to determine whether they intersected. We will not go into details, but only note that for the item set of all six items, each item was involved in several significant violations of the assumption of nonintersecting ISRFs. This assumption thus was not tenable here. This need not worry us, however, because we are mainly interested in whether the actions have the same or different orderings based on group means across different labor union divisions. Table 8.3 provides the item means within all nine divisions. Because group sizes are small, we consider only the remarkable features of the orderings and ignore small differences between orderings. It is then clear that Strike is by far the least favored means of action in each of the nine divisions, with a mean ranging from 1.1 (Publishing Industry) to 1.6 (Housing Construction). Note that the last mean is even smaller than any of the other means in Table 8.3. Notwithstanding that the divisions are very different in profession as well as in other characteristics (Van der Veen, 1992), they all seem to agree on a relatively negative attitude toward the forceful action of a strike. Less pronounced but also remarkable is that all nine divisions share the same relatively positive attitude toward protest meetings as a means to apply pressure in a conflict between the employer and the employees. Here the mean ranges from 2.5 to 2.8, and 2.5 is exceeded by only three means of 2.6 in the other rows of the table. Within divisions, the other four action mean scores are clustered within small ranges. The conclusion is that there is a consistent picture across the nine divisions, with a strike being the least favored means of action, a protest meeting the most favored means of action, and the other four means of action clustered between them.

Discussion

The context of an investigation greatly influences the interpretation of its results. Van der Veen (1992) was interested in possible differences in the ordering of means of action among union divisions, and in explaining such differences with background variables. In another context, however, say, where different orderings of the items from a word recognition test are found in different ethnic groups, this result might have been taken as evidence of item bias or differential item functioning. An important difference between Van der Veen's study and the word recognition example is that the former is aimed at understanding differences in attitude between groups, whereas the latter would be more typical of individual ability testing with an emphasis on diagnosis and possibly educational selection or job selection. In that case, equal opportunity is an important issue and may be related to possible group differences.

We finish this chapter by noting that the theory and methods of NIRT for polytomous items are more complex than those for dichotomous items. This is also true for parametric models. The simpler dichotomous models were developed first, and after enough knowledge was acquired, more complex models were developed for polytomous items and for continuous data such as response times (see Van der Linden & Hambleton, 1997, for an overview). It seems safe to conclude, however, that despite the ever-ongoing need for more theoretical development, NIRT for dichotomous items is now a mature member of the IRT family. More work needs to be done for polytomous items, but much is already known, and in our view, the availability of this knowledge and of the MSP software with its many possibilities for scale and item diagnosis also justifies the application of polytomous NIRT to practical test and questionnaire construction.

Additional Reading

Polytomous NIRT in combination with group structure was used, for example, for transitive reasoning data (Verweij, Sijtsma, & Koops, 1999) and for quality of life of cancer patients (Ringdal et al., 1999). The MSP manual (Molenaar & Sijtsma, 2000) discusses many aspects of polytomous NIRT, with the industrial malodor data as an example. For other applications with either dichotomous or polytomous items, see Appendix 1.

Exercises for Chapter 8

8.1. Summarize in at most half a page the main outcome of the analysis of the labor union data described in this chapter.

8.2. Consider the distribution of the restscore $(R_{(1)})$ in Table 8.2.
   a. Use the formula of the variance to explain why this distribution has a higher variance than a corresponding bell-shaped distribution.
   b. Which distribution is expected to have the highest total score reliability, the distribution in Table 8.2 or a bell-shaped distribution, which is also based on two four-category items?

8.3. a. Which of the item score distributions in Table 8.2 is unimodal?
   b. And which distribution is skewed?
   c. How would you, in general, expect the item score distributions to change when $R_{(1)}$ increases?

8.4. Consider the item mean column in Table 8.2; the means are mean scores on Item 1 in each of the restscore groups, to be denoted as a conditional expectation $E(X_1|R_{(1)})$.
   a. Check that each conditional item mean is the sum of the three probabilities $P(X_1 \geq h|R_{(1)})$ displayed in the last three columns.
   b. Prove for item scores $h = 0, 1, \ldots, m$ (Chapter 7) that, in general,

$$E(X_i|R_{(i)}) = \sum_{h=1}^{m} P(X_i \geq h|R_{(i)}).$$
(8.1)

Answers to Exercises for Chapter 8

8.1. The answer can best be checked by rereading the appropriate section.

8.2. a. For a population with n observations, the variance of variable R with mean μ_R is defined as

$$\sigma^2(R) = \frac{1}{n} \sum_{v=1}^{n} (R_v - \mu_R)^2.$$
(8.2)

If R has a bell-shaped distribution, most deviation scores are small, values far from the mean have low frequencies, and the mean squared deviation, which is $\sigma^2(R)$, is low. For the distribution in Table 8.2, most deviation scores are large; as a result, $\sigma^2(R)$ is high.

    b. Reliability often increases with total score variance: Thus, the distribution of Table 8.2 probably has the highest reliability.

8.3. a. For $R_{(1)} = 0$, the distribution has two modes; all other distributions have one mode.

    b. With only four score categories, check whether item scores pile up at the left or at the right, for example, by joining the lowest scores (0, 1) and the highest scores (2, 3). After joining, it is easy to see that for $R_{(1)} = 0$, scores are evenly distributed; for $R_{(1)} = 1$, 2, scores tend to pile up at the left; and for $R_{(1)} = 3 - 6$, scores tend to pile up at the right.

    c. With increasing $R_{(1)}$, modes and means are expected to shift further to the right.

8.4. a. $R_{(1)} = 0$: $1.61 = 0.77 + 0.49 + 0.35$; and so forth.

    b. In most introductory statistics textbooks, you will find that the expectation of a discrete variable X is $E(X) = \Sigma_x x P(X = x)$. Generalizing to a conditional expectation, we have, in our case,

$$E(X_i | R_{(i)}) = \sum_h h P(X_i = h | R_{(i)}). \tag{8.3}$$

As you can see, on the right the probability of item score h ($X_i = h$) appears h times. Expanding for $m = 4$ answer categories, we have

$$E(X_i | R_{(i)}) = P(X_i = 1 | R_{(i)}) + 2P(X_i = 2 | R_{(i)}) + 3P(X_i = 3 | R_{(i)}) \tag{8.4}$$

Now, writing

$$P(X_i \geq 1 | R_{(i)}) = P(X_i = 1 | R_{(i)}) + P(X_i = 2 | R_{(i)}) + P(X_i = 3 | R_{(i)});$$

$$P(X_i \geq 2 | R_{(i)}) = P(X_i = 2 | R_{(i)}) + P(X_i = 3 | R_{(i)}); \tag{8.5}$$

$$P(X_i \geq 3 | R_{(i)}) = P(X_i = 3 | R_{(i)}),$$

it can be seen that all terms appearing on the right of Equation 8.4 have been used. From this result, it follows that

$$E(X_i | R_{(i)}) = \sum_{h=1}^{3} P(X_i \geq h | R_{(i)}). \tag{8.6}$$

Obviously, this be generalized to any number of answer categories.

Appendix 1: Overview of Latent Traits Measured Successfully With a Nonparametric IRT Model

The present book discusses a class of simple nonparametric IRT models for which estimates are easily obtained from straightforward counts and whose fit to data is relatively easily investigated. Although more precise knowledge can be gained if one of the more advanced parametric models fits the data, our simple class of nonparametric IRT models has been successfully used in several domains during the more than 30 years since they were first proposed by Mokken (1971). In many of these applications the measurement of persons was the main goal, but in several others it was important to demonstrate that the difficulty order of the items was invariant across subgroups or that it conformed to predictions from a - substantive theory. The selected list of latent traits measured with a nonparametric IRT model serves to illustrate how very different substantive domains may successfully use the same measurement model.

- Perceived political influence of the interviewee (Daudt, van der Maesen, & Mokken, 1996; Mokken, 1971)
- Acquisition of knowledge and insight by young children, as a validation of the Piagetian theories in developmental psychology (Kingma, 1984; Kingma & Loth, 1985; Kingma & TenVergert, 1985; Verweij, Sijtsma, & Koops, 1996, 1999)
- Machiavelism as an attitude in politics (Henning & Six, 1977)
- Affective evaluation of different roles for men and women (Clason, 1977)
- Support for resolutions at the United Nations General Assembly (Stokman, 1977)
- Trustworthiness of the inhabitants of 13 countries (Middel & Van Schuur, 1981)
- Complaints about sleep quality (De Vries-Griever & Meijman, 1987; Meijman, Thunnissen, & De Vries-Griever, 1990)
- Attitude toward abortion (Gillespie, TenVergert, & Kingma, 1987, 1988)
- Nonverbal subtests of intelligence (Laros & Tellegen, 1991)
- Social disability of psychiatric patients (De Jong & Molenaar, 1987; Kraaikamp, 1992)

- Tiredness from workload (Meijman, 1991; Koorn, Veldman, Molenaar, & Mulder, 1995)
- Feelings of loneliness (De Jong-Gierveld & Kamphuis, 1985; Moorer & Suurmeyer, 1993)
- Quality of life of cancer patients (Ringdal & Ringdal, 1993)
- Genital sensations and body image (Van der Wiel et al., 1995)
- Handicaps of elderly persons (Suurmeyer, Doeglas, Moum, Briançon, Krol, Sanderman, Guillemin, Bjelle, & Van den Heuvel, 1994; Kempen, Miedema, Ormel, & Molenaar, 1996; Roorda, Roebroeck, Lankhorst, Van Tilburg, & Bouter, 1996)
- Handicaps of victims of a cerebrovascular accident (De Kort, 1996)
- Subjective perspective on time management (Koolhaas, Sijtsma, & Witjas, 1992)
- Verb constructions in children's language acquisition (Bol, 1994)
- Recency, Frequency and Monetary value of purchase applied to market segmentation (Paas, 1998, 1999)
- Numerical inductive reasoning (Rivas Moya, 1999)
- Inductive reasoning in visual and verbal tasks (De Koning, Sijtsma, & Hamers, 2001)
- Political participation, and wealth per household in developing countries (Zinn, Henderson, Nystuen, & Drake, 1992).

Appendix 2: List of Acronyms
(in alphabetical order)

AISP: automated item selection procedure
DIF: differential item functioning
DMM: double monotonicity model
IIO: invariant item ordering
IRF: item response function
IRT: item response theory
ISRF: item step response function
MHM: monotone homogeneity model
MSP: Mokken Scale analysis for Polytomous items
NIRT: nonparametric item response theory
1PLM: 1-parameter logistic model
3PLM: 3-parameter logistic model
2PLM: 2-parameter logistic model

Appendix 3: The MSP Software Package

MSP5 for Windows is a 32-bit program that can be ordered, together with a 104 page manual, from:

ProGAMMA BV info@gamma.rug.nl
P.O. Box 841 www.gamma.rug.nl
9700 AV Groningen phone + +31 50 577 1811
The Netherlands fax + +31 50 577 8831

A free demo version is available at the "download demo" section of ProGAMMA's Web site. You can find an overview of new software and new releases at: http://www.gamma.rug.nl/newsoftfr.html

MSP requires at least a Pentium processor with a minimum of 16 MB RAM, Microsoft Windows 95, 98, or NT, and a minimum of 5MB free disk space to install and run the program.

Appendix 4: Proof of Theorem 4.1

This proof is adapted from Mokken (1971, pp. 119, 130). Let $G(\theta)$ denote the cumulative distribution of θ across persons. Then the overall popularity of item i is given by

$$P_i = P(X_i = 1) = \int P_i(\theta) \, dG(\theta). \tag{A4.1}$$

By local independence, the probability of scoring 1 on two items i and j for a given θ is

$$P_i(\theta) \, P_j(\theta), \tag{A4.2}$$

and the overall probability of this event is

$$P_{ij} = P(X_i = 1, X_j = 1) = \int P_i(\theta) \, P_j(\theta) \, dG(\theta). \tag{A4.3}$$

Because the item scores can be only 0 or 1, one has

$$Cov(X_i, X_j) = P_{ij} - P_i P_j. \tag{A4.4}$$

In order to show that this expression is nonnegative, we first have to rewrite it in a more complicated way. In the first step, we apply Equations A4.1 and A4.3 and add a factor $\int dG(\theta)$. Note that this is equal to 1 because it is the total probability mass of the distribution of θ. Hence,

$$Cov(X_i, X_j) = \int dG(\theta) \times \int P_i(\theta) P_j(\theta) \, dG(\theta)$$

$$- \int P_i(\theta) \, dG(\theta) \times \int P_j(\theta) \, dG(\theta). \tag{A4.5}$$

Next, we use the property that any integral with integration variable θ does not change value when θ is replaced by η. In each product of two integrals, we do so for one of the factors. Moreover, it turns out that for later simplification we must first split the expression into two halves and take a different θ to η replacement in each half. This leads to

$$Cov = \frac{1}{2} \int dG(\eta) \times \int P_i(\theta)\, P_j(\theta)\, dG(\theta)$$

$$- \frac{1}{2} \int P_i(\theta)\, dG(\theta) \times \int P_j(\eta)\, dG(\eta)$$

$$+ \frac{1}{2} \int dG(\theta) \times \int P_i(\eta)\, P_j(\eta)\, dG(\eta)$$

$$- \frac{1}{2} \int P_i(\eta)\, dG(\eta) \times \int P_j(\theta)\, dG(\theta) \qquad \text{(A4.6)}$$

$$= \frac{1}{2} \int\int [P_i(\theta)\, P_j(\theta) - P_i(\theta)\, P_j(\eta)$$

$$+ P_i(\eta)\, P_j(\eta) - P_i(\eta)\, P_j(\theta)]\, dG(\theta) dG(\eta)$$

$$= \frac{1}{2} \int\int [P_i(\theta) - P_i(\eta)]\, [P_j(\theta) - P_j(\eta)]\, dG(\theta) dG(\eta).$$

In the part of the (θ, η) plane where $\theta > \eta$, both factors between square brackets are nonnegative because of the monotonicity of the IRFs. In the remaining part, both factors are nonpositive and their product is again nonnegative. The integral of a nonnegative integrand cannot be negative, so the covariance of the item scores cannot be negative. This completes the proof.

References

Baker, F. B. (1992). *Item response theory*. New York: Marcel Dekker.

Bleichrodt, N., Drenth, P. J. D., Zaal, J. N., & Resing, W. C. M. (1985). *Revisie Amsterdamse Kinderintelligentie Test (RAKIT)* [Revision of the Amsterdam Child Intelligence Test]. Lisse: Swets & Zeitlinger.

Bol, G. W. (1994). Implicational scaling in child language acquisition: The order of production of Dutch verb constructions. In M. Verrips & F. Wijnen (Eds.), *Papers from the Dutch-German colloquium on language acquisition* (Amsterdam Series in Child Language Development). Amsterdam: University of Amsterdam, Institute for General Linguistics.

Bolt, D. M. (2001). Conditional covariance-based representation of multidimensional test structure. *Applied Psychological Measurement, 25*, 244–257.

Boomsma, A., Van Duijn, M. A. J., & Snijders, T. A. B. (Eds.), (2001). *Essays on item response theory*. New York: Springer.

Carroll, J. B. (1961). The nature of the data, or how to choose a correlation coefficient. *Psychometrika, 26*, 347–372.

Cavalini, P. M. (1992). *It's an ill wind that brings no good. Studies on odour annoyance and the dispersion of odorant concentrations from industries*. Ph.D. thesis, University of Groningen, The Netherlands.

Chang, H.-H., Mazzeo, J., & Roussos, L. (1996). Detecting DIF for polytomously scored items: An adaption of the SIBTEST procedure. *Journal of Educational Measurement, 33*, 333–353.

Clason, C. E. (1977). *Beroepsarbeid door gehuwde vrouwen* [Professional work by married women]. Ph.D. thesis, University of Groningen, The Netherlands.

Cliff, N. (1977). A theory of consistency of ordering generalizable to tailored testing. *Psychometrika, 42*, 375–401.

Cronbach, L. J. (1951). Coefficient alpha and the internal structure of tests. *Psychometrika, 16*, 297–334.

Croon, M. A. (1991). Investigating Mokken scalability of dichotomous items by means of ordinal latent class analysis. *British Journal of Mathematical and Statistical Psychology, 44*, 315–331.

Cudeck, R. (1980). A comparative study of indices for internal consistency. *Journal of Educational Measurement, 17*, 117–130.

Daudt, H., Van der Maesen, C. E., & Mokken, R. J. (1996). Political efficacy: A further exploration. *Acta Politica, 31*, 350–371.

De Jong, A., & Molenaar, I. W. (1987). An application of Mokken's model for stochastic cumulative scaling in psychiatric research. *Journal of Psychiatric Research, 21*, 137–149.

De Jong-Gierveld, J., & Kamphuis, F. (1985). The development of a Rasch-type loneliness scale. *Applied Psychological Measurement, 9*, 289–299.

De Koning, E., Sijtsma, K., & Hamers, J. H. M. (2001). *Comparison of four IRT models when analyzing two tests for inductive reasoning*. Manuscript submitted for publication.

De Kort, P. L. M. (1996). *Neglect. Een klinisch onderzoek naar halfzijdige verwaarlozing bij patiënten met een cerebrale bloeding of infarct* [Neglect. A clinical study of semilateral neglect in patients with a cerebrovascular accident]. Ph.D. thesis, University of Groningen, The Netherlands.

De Vries-Griever, A. H. G., & Meijman, T. F. (1987). The impact of abnormal hours of work on various modes of information processing: A process model of human costs of performance. *Ergonomics, 30,* 1287–1299.

Douglas, J., & Cohen, A. (2001). Nonparametric item response function estimation for assessing parametric model fit. *Applied Psychological Measurement, 25,* 234–243.

Douglas, J., Kim, H. R., Habing, B., & Gao, F. (1998). Investigating local dependence with conditional covariance functions. *Journal of Educational and Behavioral Statistics, 23,* 129–151.

Ellis, J. L., & Junker, B. W. (1997). Tail-measurability in monotone latent variable models. *Psychometrika, 62,* 495–523.

Emons, W. H. M., Meijer, R. R., & Sijtsma, K. (in press). Comparing simulated and theoretical sampling distributions of the U3 person-fit statistic. *Applied Psychological Measurement.*

Fischer, G. H., & Molenaar, I. W. (Eds.). (1995). *Rasch models. Foundations, recent developments, and applications.* New York: Springer.

Gillespie, M., TenVergert, E. M., & Kingma, J. (1987). Using Mokken scale analysis to develop unidimensional scales. *Quality & Quantity, 21,* 393–408.

Gillespie, M., TenVergert, E. M., & Kingma, J. (1988). Secular trends in abortion attitudes: 1975–1980–1985. *The Journal of Psychology, 122,* 323–341.

Grayson, D. A. (1988). Two-group classification in latent trait theory: Scores with monotone likelihood ratio. *Psychometrika, 53,* 383–392.

Guilford, J. P. (1954). *Psychometric methods.* New York: McGraw-Hill.

Guttman, L. (1945). A basis for analyzing test-retest reliability. *Psychometrika, 10,* 255–282.

Habing, B. (2001). Nonparametric regression and the parametric bootstrap for local dependence assessment. *Applied Psychological Measurement, 25,* 221–233.

Hambleton, R. K., & Swaminathan, H. (1985). *Item response theory. Principles and applications.* Boston: Kluwer Nijhoff.

Hambleton, R. K., Swaminathan, H., & Rogers, H. J. (1991). *Fundamentals of items response theory.* Newbury Park, CA: Sage.

Hemker, B. T., Sijtsma, K., & Molenaar, I. W. (1995). Selection of unidimensional scales from a multidimensional itembank in the polytomous Mokken IRT model. *Applied Psychological Measurement, 19,* 337–352.

Hemker, B. T., Sijtsma, K., Molenaar, I. W., & Junker, B. W. (1996). Polytomous IRT models and monotone likelihood ratio of the total score. *Psychometrika, 61,* 679–693.

Hemker, B. T., Sijtsma, K., Molenaar, I. W., & Junker, B. W. (1997). Stochastic ordering using the latent trait and the sum score in polytomous IRT models. *Psychometrika, 62,* 331–347.

Hemker, B. T., Van der Ark, L. A., & Sijtsma, K. (2001). On measurement properties of continuation ratio models. *Psychometrika, 66,* 487–506.

Henning, H. J., & Six, B. (1977). Konstruktion einer Machiavellismus-Skala [Construction of a Machiavelism scale]. *Zeitschrift für Sozialpsychologie, 8,* 185–198.

Hoijtink, H., & Molenaar, I. W. (1997). A multidimensional item response model: Constrained latent class analysis using the Gibbs sampler and posterior predictive checks. *Psychometrika, 62,* 171–189.

Holland, P. W. (1990). On the sampling theory foundations of item response theory models. *Psychometrika, 55*, 577–601.

Holland, P. W., & Rosenbaum, P. R. (1986). Conditional association and unidimensionality in monotone latent variable models. *The Annals of Statistics, 14*, 1523–1543.

Holland, P. W., & Wainer, H. (Eds.) (1993). *Differential item functioning*. Hillsdale, NJ: Lawrence Erlbaum.

Horst, P. (1953). Correcting the Kuder-Richardson reliability for dispersion of item difficulties. *Psychological Bulletin, 50*, 371–374.

Huisman, M., & Molenaar, I. W. (2001). Imputation of missing scale data with item response theory. In A. Boomsma, M. A. J. van Duijn, & T. A. B. Snijders (Eds.), *Essays on item response theory* (pp. 221–244). New York: Springer.

Huynh, H. (1994). A new proof for monotone likelihood ratio for the sum of independent Bernoulli random variables. *Psychometrika, 59*, 77–79.

Jansen, P. G. W. (1981). Spezifisch objektive Messung im Falle monotoner Einstellungsitems [Specifically objective measurement of monotone attitude items]. *Zeitschrift für Sozialpsychologie, 12*, 24–41.

Jansen, P. G. W. (1982a). De onbruikbaarheid van Mokkenschaalanalyse [On the uselessness of Mokken scale analysis]. *Tijdschrift voor Onderwijsresearch, 7*, 11–24.

Jansen, P. W. G. (1982b). Homogenitätsmessung mit Hilfe des Koeffizienten H von Loevinger: Eine kritische Diskussion [Measuring homogeneity by means of Loevinger's coefficient H: A critical discussion]. *Psychologische Beiträge, 24*, 96–105.

Jansen, P. G. W. (1983). *Rasch analysis of attitudinal data*. Ph.D. thesis, Catholic University of Nijmegen, The Netherlands.

Jansen, P. G. W., Roskam, E. E. Ch. I., & Van den Wollenberg, A. L. (1982). De Mokkenschaal gewogen. (Weighing the Mokken scaling procedure). *Tijdschrift voor Onderwijsresearch, 7*, 31–42.

Jansen, P. G. W., Roskam, E. E. Ch. I., & Van den Wollenberg, A. L. (1984). Discussion on the usefulness of the Mokken procedure for nonparametric scaling. *Psychologische Beiträge, 26*, 722–735.

Junker, B. W. (1991). Essential independence and likelihood-based ability estimation for polytomous items. *Psychometrika, 56*, 255–278.

Junker, B. W. (1993). Conditional association, essential independence and monotone unidimensional item response models. *The Annals of Statistics, 21*, 1359–1378.

Junker, B. W. (2001). On the interplay between nonparametric and parametric IRT, with some thoughts about the future. In A. Boomsma, M. A. J. van Duijn, & T. A. B. Snijders (Eds.), *Essays on item response theory* (pp. 247–276). New York: Springer.

Junker, B. W., & Sijtsma, K. (2000). Latent and manifest monotonicity in item response models. *Applied Psychological Measurement, 24*, 65–81.

Junker, B. W., & Sijtsma, K. (2001). Nonparametric item response theory in action: An overview of the special issue. *Applied Psychological Measurement, 25*, 211–220.

Kelderman, H., & Rijkes, C. P. M. (1994). Loglinear multidimensional IRT models for polytomously scored items. *Psychometrika, 59*, 149–176.

Kempen, G. I. J. M., Miedema, I., Ormel, J., & Molenaar, W. (1996). The assessment of disability with the Groningen Activity Restriction Scale. Conceptional framework and psychometric properties. *Social Science & Medicine, 11*, 1601–1610.

Kingma, J. (1984). A comparison of four methods of scaling for the acquisition of early number concept. *The Journal of General Psychology, 110*, 23–45.

Kingma, J., & Loth, F. L. (1985). The validation of a developmental scale for seriation. *Educational and Psychological Measurement, 45*, 321–328.

Kingma, J., & Ten Vergert, E. M. (1985). A nonparametric scale analysis of the development of conservation. *Applied Psychological Measurement, 9*, 375–387.

Koolhaas, M. J., Sijtsma, K., & Witjas, R. (1992). Tijdperspectieven in time management trainingen. Enkele psychometrische aspecten van een vragenlijst [Time perspectives in time management training. Some psychometric aspects of a questionnaire]. *Gedrag en Organisatie, 5*, 94–105.

Koorn, R., Veldman, J. B. P., Molenaar, I. W., & Mulder, L. J. M. (1995). Werken aan beeldschermen: De ontwikkeling van een visuele klachtenlijst met behulp van een Mokkenanalyse [Working at computer screens: The development of a list of visual complaints via a Mokken scale analysis]. *Gedrag en Organisatie, 8*, 318–331.

Kraaikamp, H. J. M. (1992). *Moeilijke rollen. Psychometrisch onderzoek naar de betrouwbaarheid en validiteit van de Groningse Sociale Beperkingenschaal bij psychiatrische patienten* [Difficult roles. Psychometric research of the reliability and validity of the Groningen Social Restriction Scale for psychiatric patients]. Ph.D. thesis, University of Groningen, The Netherlands.

Laros, J. A., & Tellegen, P. J. (1991). *Construction and validation of the SON-R 5 1/2 – 17, the Snijders-Oomen non-verbal intelligence test.* Groningen, The Netherlands: Wolters-Noordhoff.

Loevinger, J. (1947). A systematic approach to the construction and evaluation of tests of ability. *Psychological Monographs, 61*, No. 4.

Loevinger, J. (1948). The technique of homogeneous tests compared with some aspects of "scale analysis" and factor analysis. *Psychological Bulletin, 45*, 507–530.

Lord, F. M. (1980). *Applications of item response theory to practical testing problems.* Hillsdale, NJ: Lawrence Erlbaum.

Meijer, R. R. (Ed.) (1996). Person-fit research: Theory and application [Special issue]. *Applied Measurement in Education, 9*(1).

Meijer, R. R., & Sijtsma, K. (2001). Methodology review: Evaluating person fit. *Applied Psychological Measurement, 25*, 107–135.

Meijer, R. R., Sijtsma, K., & Smid, N. G. (1990). Theoretical and empirical comparison of the Mokken and the Rasch approach to IRT. *Applied Psychological Measurement, 14*, 283–298.

Meijman, T. F. (1991). *Over Vermoeidheid. Arbeidspsychologische studies naar de beleving van belastingseffecten* [About tiredness. Labor psychology studies of the experienced mental load]. Ph.D. thesis, University of Groningen, The Netherlands.

Meijman, T. F., Thunnissen, M. J., & De Vries-Griever, A. G. H. (1990). The after-effects of a prolonged period of day-sleep on subjective sleep quality. *Work & Stress, 4*, 65–70.

Middel, B. P., & Van Schuur, W. (1981). Background characteristics, attitudes towards the European Community and towards Dutch politics, of delegates from CDA, D'66, PvdA, and VVD. *Acta Politica, 16*, 241–263.

Mokken, R. J. (1971). *A theory and procedure of scale analysis.* The Hague: Mouton/Berlin: De Gruyter.

Mokken, R. J. (1997). Nonparametric models for dichotomous responses. In W. J. van der Linden & R. K. Hambleton (Eds.), *Handbook of modern item response theory* (pp. 351–367). New York: Springer.

Mokken, R. J., & Lewis, C. (1982). A nonparametric approach to the analysis of dichotomous item responses. *Applied Psychological Measurement, 6*, 417–430.

Mokken, R. J., Lewis, C., & Sijtsma, K. (1986). Rejoinder to "The Mokken scale: A critical discussion." *Applied Psychological Measurement, 10*, 279–285.

Molenaar, W. (1970). Approximations to the Poisson, binomial and hypergeometric distributions. *Mathematical Centre Tracts, No. 31*, Amsterdam.

Molenaar, I. W. (1982a). De beperkte bruikbaarheid van Jansens kritiek [On the limited usefulness of Jansen's criticisms]. *Tijdschrift voor Onderwijsresearch, 7*, 25–30.

Molenaar, I. W. (1982b). Een tweede weging van de Mokkenschaal [A second weighing of the Mokken scaling procedure]. *Tijdschrift voor Onderwijsresearch, 7*, 172–181.

Molenaar, I. W. (1982c). Mokken scaling revisited. *Kwantitatieve Methoden, 3*(8), 145–164.

Molenaar, I. W. (1991). A weighted Loevinger H-coefficient extending Mokken scaling to multicategory items. *Kwantitatieve Methoden, 12*(37), 97–117.

Molenaar, I. W. (1997). Nonparametric models for polytomous responses. In W. J. van der Linden & R. K. Hambleton (Eds.), *Handbook of modern item response theory* (pp. 369–380). New York: Springer.

Molenaar, I. W., & Sijtsma, K. (1988). Mokken's approach to reliability estimation extended to multicategory items. *Kwantitatieve Methoden, 9*(28), 115–126.

Molenaar, I. W., & Sijtsma, K. (2000). *User's manual MSP5 for Windows.* Groningen: iecProGAMMA.

Moorer, P., & Suurmeyer, T. P. B. M. (1993). Unidimensionality and cumulativeness of the Loneliness Scale using Mokken scale analysis for polychotomous items. *Psychological Reports, 73*, 1324–1326.

Nunnally, J. C. (1978). *Psychometric theory.* New York: McGraw-Hill.

Paas, L. J. (1998). Mokken scaling characteristic sets and acquisition patterns of durable and financial products. *Journal of Economic Psychology, 19*, 353–376.

Paas, L. J. (1999). Refining RFM-variables through Mokken scale analysis for the purpose of optimal prospect selection: Application to ownership patterns of financial products. *Journal of Market Focused Management, 3*, 275–294.

Post, W. J. (1992). *Nonparametric unfolding models. A latent structure approach.* Leiden University, The Netherlands: DSWO Press.

Post, W. J., & Snijders, T. A. B. (1993). Nonparametric unfolding models for dichotomous scaling data. *Methodika, 7*, 130–156.

Raju, N. S. (1982). On tests of homogeneity and maximum KR-20. *Educational and Psychological Measurement, 42*, 145–152.

Ramsay, J. O. (1991). Kernel smoothing approaches to nonparametric item characteristic curve estimation. *Psychometrika, 56*, 611–630.

Reckase, M. D. (1997). A linear logistic multidimensional model for dichotomous item response data. In W. J. van der Linden & R. K. Hambleton (Eds.), *Handbook of modern item response theory* (pp. 271–286). New York: Springer.

Ringdal, G. I., & Ringdal, K. (1993). Testing the EORTC Quality of Life Questionnaire on cancer patients with heterogeneous diagnosis. *Quality of Life Research, 2*, 129–140.

Ringdal, K., Ringdal, G. I., Kaasa, S., Bjordal, K., Wislöff, F., Sundström, S., & Hjermstad, M. J. (1999). Assessing the consistency of psychometric properties of the HRQoL scales within the EORTC QLQ-C30 across populations by means of the Mokken scaling model. *Quality of Life Research, 8*, 25–43.

Rivas Moya, T. (1999). *Mokken scale analysis: An application to items of numerical inductive reasoning.* Paper presented at the European Meeting of the Psychometric Society, Lueneburg, Germany.

Roorda, L. D., Roebroeck, M. E., Lankhorst, G. J., Van Tilburg, T., & Bouter, L. M. (1996). Measuring functional limitations in rising and sitting down: Development of a questionnaire. *Archives of Physical Medicine and Rehabilitation, 77,* 663–669.

Rosenbaum, P. R. (1984). Testing the conditional independence and monotonicity assumptions of item response theory. *Psychometrika, 49,* 425–435.

Rosenbaum, P. R. (1987a). Comparing item characteristic curves. *Psychometrika, 52,* 217–233.

Rosenbaum, P. R. (1987b). Probability inequalities for latent scales. *British Journal of Mathematical and Statistical Psychology, 40,* 157–168.

Roskam, E. E., Van den Wollenberg, A. L., & Jansen, P. G. W. (1986). The Mokken scale: A critical discussion. *Applied Psychological Measurement, 10,* 265–277.

Roussos, L. A., Stout, W. F., & Marden, J. (1998). Using new proximity measures with hierarchical cluster analysis to detect multidimensionality. *Journal of Educational Measurement, 35,* 1–30.

Samejima, F. (1969). Estimation of latent trait ability using a response pattern of graded scores. *Psychometrika Monograph, No. 17.*

Samejima, F. (1972). A general model for free-response data. *Psychometrika Monograph, No. 18.*

Shealy, R. T., & Stout, W. F. (1993). An item response theory model for test bias and differential item functioning. In P. W. Holland & H. Wainer (Eds.), *Differential item functioning* (pp. 197–239). Hillsdale, NJ: Lawrence Erlbaum.

Sijtsma, K. (1984). Useful nonparametric scaling: A reply to Jansen. *Psychologische Beiträge, 26,* 423–437.

Sijtsma, K. (1986). Another note on the usefulness of Mokken scaling. *Psychologische Beiträge, 28,* 425–432.

Sijtsma, K. (1988). *Contributions to Mokken's nonparametric item response theory.* Amsterdam: Free University Press.

Sijtsma, K. (1998). Methodology review: Nonparametric IRT approaches to the analysis of dichotomous item scores. *Applied Psychological Measurement, 22,* 3–32.

Sijtsma, K. (2001). Developments in measurement of persons and items by means of item response models. *Behaviormetrika, 28,* 65–94.

Sijtsma, K., Debets, P., & Molenaar, I. W. (1990). Mokken scale analysis for polychotomous items: Theory, a computer program and an empirical application. *Quality & Quantity, 24,* 173–188.

Sijtsma, K., & Hemker, B. T. (1998). Nonparametric polytomous IRT models for invariant item ordering, with results for parametric models. *Psychometrika, 63,* 183–200.

Sijtsma, K., & Hemker, B. T. (2000). A taxonomy of IRT models for ordering persons and items using simple sum scores. *Journal of Educational and Behavioral Statistics, 25,* 391–415.

Sijtsma, K., & Junker, B. W. (1996). A survey of theory and methods of invariant item ordering. *British Journal of Mathematical and Statistical Psychology, 49,* 79–105.

Sijtsma, K., & Junker, B. W. (1997). Invariant item ordering of transitive reasoning tasks. In J. Rost & R. Langeheine (Eds.), *Applications of latent trait and latent class models in the social sciences* (pp. 100–110). Münster, Germany: Waxmann Verlag.

Sijtsma, K., & Meijer, R. R. (1992). A method for investigating the intersection of item response functions in Mokken's nonparametric IRT model. *Applied Psychological Measurement, 16,* 149–157.

Sijtsma, K., & Molenaar, I. W. (1987). Reliability of test scores in nonparametric item response theory. *Psychometrika, 52,* 79–97.

Sijtsma, K., & Prins, P. M. (1986). Itemselectie in het Mokken model [Item selection in the Mokken model]. *Tijdschrift voor Onderwijsresearch, 11*, 121–129.

Sijtsma, K., & Van der Ark, L. A. (2001). Progress in NIRT analysis of polytomous item scores: Dilemmas and practical solutions. In A. Boomsma, M. A. J. van Duijn, & T. A. B. Snijders (Eds.), *Essays on item response theory* (pp. 297–318). New York: Springer.

Snijders, T. A. B. (2001). Two-level nonparametric scaling for dichotomous data. In A. Boomsma, M. A. J. van Duijn, & T. A. B. Snijders (Eds.), *Essays on item response theory* (pp. 319–338). New York: Springer.

Stokman, F. N. (1977). *Roll calls and sponsorship: A methodological analysis of Third World group formation in the United Nations.* Leiden: Sijthoff.

Stout, W. F. (1987). A nonparametric approach for assessing latent trait unidimensionality. *Psychometrika, 52*, 589–617.

Stout, W. F. (1990). A new item response theory modeling approach with applications to unidimensionality assessment and ability estimation. *Psychometrika, 55*, 293–325.

Suurmeyer, T. P. B. M., Doeglas, D. M., Moum, T., Briançon, S., Krol, B., Sanderman, R., Guillemin, F., Bjelle, A., & Van den Heuvel, W. J. A. (1994). The Groningen Activity Restriction Scale for Measuring Disability: Its utility in international comparisons. *American Journal of Public Health, 84*, 1270–1273.

Terwilliger, J. S., & Lele, K. T. (1979). Some relationships among internal consistency, reproducibility and homogeneity. *Journal of Educational Measurement, 16*, 101–108.

Van Abswoude, A. A. H., Van der Ark, L. A., & Sijtsma, K. (2001). *A comparative study on test dimensionality assessment procedures under nonparametric IRT models.* Manuscript submitted for publication.

Van der Ark, L. A. (2000). *Practical consequences of stochastic ordering of the latent trait under various polytomous IRT models.* Manuscript submitted for publication.

Van der Ark, L. A. (2001). Relationships and properties of polytomous item response theory models. *Applied Psychological Measurement, 25*, 273–282.

Van der Linden, W. J., & Hambleton, R. K. (Eds.). (1997). *Handbook of modern item response theory.* New York: Springer.

Van der Veen, G. (1992). *Principes in praktijk. CNV-leden over collectieve acties* [Principles into practice. Labor union members on means of political pressure]. Kampen: Kok.

Van der Wiel, H. B. M., Weijmar Schultz, W. C. M., Molenaar, I. W., Vennix, P., Beens, H., & Vessies, D. (1995). Zelfbeoordeling van genitale sensaties en lichaamsperceptie bij vrouwen; de constructie van twee vragenlijsten [Self-rating of genital sensations and body perception of women: The construction of two questionnaires]. *Tijdschrift voor Seksuologie, 19*, 119–131.

Van Engelenburg, G. (1997). *On psychometric models for polytomous items with ordered categories within the framework of item response theory.* Ph.D. thesis, University of Amsterdam, The Netherlands.

Van Schuur, W. H. (1984). *Structure in political beliefs. A new model for stochastic unfolding with application to European party activists.* Amsterdam: CT Press.

Vermunt, J. K. (2001). The use of restricted latent class models for defining and testing nonparametric and parametric item response theory models. *Applied Psychological Measurement, 25*, 283–294.

Verweij, A. C. (1994). *Scaling transitive inference in 7–12 year old children.* Ph.D. Thesis, Free University of Amsterdam, The Netherlands.

Verweij, A. C., Sijtsma, K., & Koops, W. (1996). A Mokken scale for transitive reasoning suited for longitudinal research. *International Journal of Behavioral Development*, *19*, 219–238.

Verweij, A. C., Sijtsma, K., & Koops, W. (1999). An ordinal scale for transitive reasoning by means of a deductive strategy. *International Journal of Behavioral Development*, *23*, 241–264.

Wechsler, D. (1999). *WISC manual*. San Antonio, TX: Psychological Corporation.

Zhang, J., & Stout, W. F. (1999). The theoretical DETECT index of dimensionality and its application to approximate simple structure. *Psychometrika*, *64*, 213–249.

Zinn, F. D., Henderson, D. A., Nystuen, J. D., & Drake, W. D. (1992). A stochastic cumulative scaling method applied to measuring wealth in Indonesian villages. *Environment and Planning A*, *24*, 1155–1166.

Index

About the Authors

Klaas Sijtsma (PhD, University of Groningen) is a Full Professor of Methodology of Psychological Research at Tilburg University, Tilburg, The Netherlands. He is a member of the Psychometric Society, and the Netherlands Society for Statistics and Operations Research. He is on the Board of Trustees of CITO National Institute of Educational Measurement, and on the Editorial Board of *Applied Psychological Measurement*. His research interests are psychometric research in item response theory, in particular the cognitive modeling of psychological test data, nonparametric item response models, person-fit methods, and polytomous item response models; missing data and outlier problems in factor analysis and test theory; and test and questionnaire construction. Dr. Sijtsma's work has appeared in *Psychometrika, Journal of Educational and Behavioral Statistics, Applied Psychological Measurement*, and the *British Journal of Mathematical and Statistical Psychology*.

Ivo W. Molenaar (PhD, University of Amsterdam) is Emeritus Professor of Statistics and Measurement, University of Groningen, The Netherlands. He is a past president of the Psychometric Society and a former editor of *Psychometrika*, a member of the International Statistical Institute, and a past president of the VvS Netherlands Society for Statistics and Operations Research. His research is in psychometric models for the measurement of mental abilities and attitudes, in particular the Rasch model and the Mokken model in item response theory; Baysian methods, particularly prior elicitation and robustness of model choice; and behavior studies of the users of statistical software. Dr. Molenaar is coeditor (with G. H. Fischer) of *Rasch Models: Foundations, Recent Developments and Applications* (1995) and has published several book chapters. His work also appears in *Psychometrika* and *Computational Statistics and Data Analysis*, among other journals.

Printed in the United States
By Bookmasters

Printed in the United States
By Bookmasters